AF404724

A Classical Thermodynamics Toolkit

Online at: https://doi.org/10.1088/978-0-7503-6029-6

A Classical Thermodynamics Toolkit

Srinivas Vanapalli
*Faculty of Science and Technology, University of Twente (Netherlands),
Enschede, The Netherlands*

IOP Publishing, Bristol, UK

ISBN   978-0-7503-6029-6 (ebook)
ISBN   978-0-7503-6027-2 (print)
ISBN   978-0-7503-6030-2 (myPrint)
ISBN   978-0-7503-6028-9 (mobi)

DOI   10.1088/978-0-7503-6029-6

Version: 20251201

IOP ebooks

British Library Cataloguing-in-Publication Data: A catalogue record for this book is available from the British Library.

Published by IOP Publishing, wholly owned by The Institute of Physics, London

IOP Publishing, No.2 The Distillery, Glassfields, Avon Street, Bristol, BS2 0GR, UK

US Office: IOP Publishing, Inc., 190 North Independence Mall West, Suite 601, Philadelphia, PA 19106, USA

*To the memory of **Professor Miko Elwenspoek**, my PhD supervisor and lifelong inspiration.*

His unwavering encouragement to stay curious, seek diverse perspectives, and explain complex ideas with clarity has shaped not only this book but also the way I approach science. To me, he was like a Richard Feynman – but one I had the privilege to know, learn from, and walk alongside.

Contents

Preface

The world has changed. We no longer live in the age of steam engines, and even internal combustion engines are fading into the background. Yet, many thermodynamics textbooks continue to reflect the priorities of that era. They are often written for mechanical engineers or physicists, each with their own framing of the subject—and understandably so. But as we move deeper into the 21st century, we must prepare scientists and engineers for a new reality: one that is inherently multidisciplinary.

This book is written with that future in mind.

The idea of making thermodynamics accessible across disciplines was first instilled in me through the Advanced Technology program at the University of Twente—a visionary curriculum launched over two decades ago by my late PhD supervisor, Professor Miko Elwenspoek. He always encouraged me to keep an open mind, to seek different perspectives, and to explain things simply. That spirit runs through every chapter of this book.

Whether you specialize in mechatronics, robotics, semiconductors, chemical processes, aerospace, or applied physics, a foundational understanding of thermodynamics is essential. But thermodynamics is often seen as opaque, filled with abstract definitions and mysterious quantities—none more so than entropy. Many textbooks also fail to clearly distinguish the *system* from its *surroundings*, or they develop ideas in ways that obscure their practical relevance. This book attempts to break through those barriers. It is written with no inherent assumptions, built from the ground up to foster clear, intuitive understanding. My goal is to give every reader—whether a student or a practicing scientist—a platform from which they can take off in any specialized direction.

This book is for **everyone who is curious about Nature**. It requires no advanced mathematics; elementary calculus is helpful but not essential. A first-year bachelor student from any discipline can benefit from it. Practicing engineers and scientists may also find it a useful reference, particularly for revisiting concepts that are too often rushed in traditional education.

The structure of this book has been shaped by more than a decade of teaching and research. When I began teaching thermodynamics at the University of Twente 14 years ago, students would often ask me questions that I didn't always have good answers for. That pushed me to sharpen my own understanding—not just by reading, but by *working through* the ideas until they clicked. I kept notes on the questions that challenged me most, and those notes gradually evolved into this book.

Importantly, many of the insights here also stem from my research work in the lab. It was in the context of real experimental systems that I began to see how the abstract concepts of thermodynamics come to life—how energy flows, how entropy reveals itself, and how to reason about systems rigorously yet simply.

This practical grounding led me to an essential realization: thermodynamics is **universal**. It's not just about gases, engines, or power cycles. It applies just as much to electrical, magnetic, and chemical systems. I remember teaching thermodynamics to applied physics students who asked, 'Why should we learn about chemical reactions?' My answer was that even boiling water—a phase change—can be interpreted as a kind of chemical transformation. Learning fundamental thermodynamics allows us to translate insights across domains.

And yes, we often start with the classic piston–cylinder system—not because it's outdated, but because it offers an elegant, visual way to grasp foundational ideas like pressure, volume, energy, and entropy. But the tools developed in this book go far beyond that. They provide a language for understanding a wide range of natural and engineered systems, from simple heat exchange to complex technological applications.

Along the way, I've also noticed that many learners confuse thermodynamics with other fields such as heat transfer or kinetics. A helpful analogy is the game of chess: thermodynamics defines the rules—what is allowed, what is forbidden—while kinetics and heat transfer determine how fast the game unfolds. Both are important, but this book is about the rules. Clarifying such distinctions is part of what makes the subject approachable and useful.

This book is the product of many conversations, mistakes, realizations, and refinements—all of which I am grateful for. I hope it serves not only as a learning resource but also as an invitation: to look more closely, to question confidently, and to carry the spirit of curiosity into your own work, wherever it leads.

Srinivas Vanapalli
University of Twente

Acknowledgements

Although my academic background is in electrical engineering, it was Ray Radebaugh (NIST, Boulder, Colorado) who first inspired my deep interest in thermodynamics. His clarity of thought and passion for the subject left a lasting impression on me, and I am deeply grateful for that influence.

At the University of Twente, my teaching experience has been enriched by many people. I want to especially acknowledge Jelle van der Meulen, whose creative style of designing exercises—often like solving puzzles or filling in thought-provoking tables—inspired a more intuitive and engaging approach to teaching. My co-teacher Herman Hemmes has also been a consistent source of support and insight throughout the years.

This book would not have taken its current form without the help of my PhD students, Rick Spijkers and Adam Kovács, who contributed to many of the illustrations and figures. I sincerely thank them for their patience and eye for detail.

I am deeply grateful to the students of Advanced Technology and Applied Physics at the University of Twente. Their curiosity, questions, and willingness to challenge the material pushed me to refine my explanations and clarify my thinking. Much of this book was shaped by those classroom interactions.

I am also thankful to IOP Publishing for their proactive and professional support throughout the publication process. Their clear communication and attention to detail helped turn the manuscript into a well-produced final book.

Lastly, I want to express my heartfelt appreciation to my wife and son for their constant encouragement and patience during this journey. Their support made all the difference.

Author biography

Srinivas Vanapalli

Srinivas Vanapalli is a Professor of Applied Physics at the University of Twente, where he leads the Applied Thermal Sciences (ATS) group. His team is pioneering a revolution in cryogenics—rethinking traditional systems to make them energy-efficient, compact, scalable, and sustainable. They develop next-generation cryocoolers with dramatically reduced power consumption, thermal management solutions like cryogenic circulators and heat switches, and even liquid-hydrogen zero boil-off systems and modular cryogenic platforms.

Originally trained as an electrical engineer, Professor Vanapalli earned his PhD conducting cryocooler research at NIST in Boulder, Colorado. His work has since bridged fundamental science, industrial innovation, and focused entrepreneurship—collaborating with companies and developing novel cryogenic technologies for quantum computing, energy systems, and beyond.

A committed educator, he has taught thermodynamics for over 14 years, known for his clear, intuitive approach shaped by both student interactions and hands-on lab research. He co-designed problem-based exercises that bring concepts to life, and his PhD students have supported the book with technical illustrations.

Professor Vanapalli also serves the global scientific community—he is Chairman of the Cryogenics Society of Europe and Chair of the Cryogenic Engineering Conference (CEC)—working to raise professional standards and foster collaboration across academia and industry.

This book reflects his belief in universal thermodynamics: that the same principles underpin diverse systems—from engines to cryocoolers to hydrogen energy infrastructure—and that clarity and simplicity are the keys to unlocking innovation.

IOP Publishing

A Classical Thermodynamics Toolkit

Srinivas Vanapalli

Chapter 1

Introduction

This chapter introduces the foundational perspectives and principles of classical thermodynamics, focusing on the dual approach of examining systems from both microscopic and macroscopic viewpoints. In thermodynamics, we can describe the behavior of systems at a detailed molecular level or from an averaged perspective. Each approach provides valuable insights into the properties and interactions of matter in various states—solid, liquid, and gas—and how they respond to changes in energy and forces.

We explore key thermodynamic properties such as internal energy, the concept of the system and surroundings, and how systems exchange energy through heat and work. By examining simple models, such as a cube of gas or phase changes of water, we illustrate the essential processes of energy transfer. Additionally, we discuss the ideal gas law and how real gases deviate from ideal behavior under certain conditions, providing a basis for understanding the complexities of more advanced thermodynamic systems.

This chapter also introduces essential tools for studying force balance and property relationships in thermodynamics. Together, these principles set the stage for a deeper examination of thermodynamic processes and the practical applications of energy transformations, as seen in the boundary work of piston–cylinder devices. Through exercises and examples, readers will gain insight into the dynamic exchanges between energy forms and how these interactions define the structure and function of systems in thermodynamic equilibrium and beyond.

1.1 Microscopic and macroscopic viewpoints

The behavior of a system may be investigated from either a microscopic or macroscopic point of view. Let us briefly describe a system from a microscopic point of view. Consider a system consisting of a cube 25 mm on a side and containing a monatomic gas at atmospheric pressure and temperature. This volume contains approximately 10^{20} atoms. We are interested in predicting the properties of

doi:10.1088/978-0-7503-6029-6ch1

1-1

this system. There are two approaches to this problem that reduce the number of equations and variables to a few that can be computed relatively easily. One is the statistical approach, in which, based on statistical considerations and probability theory, we deal with average values for all particles under consideration. This is usually done in connection with a model of the atom under consideration. This is the approach used in the disciplines of kinetic theory and statistical mechanics.

The other approach to reducing the number of variables to a few that can be handled relatively easily involves the macroscopic point of view of classical thermodynamics. As the word macroscopic implies, we are concerned with the gross or average effects of many molecules. These effects can be perceived by our senses and measured by instruments.

However, what we really perceive, and measure, is the time-averaged influence of many molecules. For example, consider the pressure a gas exerts on the walls of its container. This pressure results from the change in momentum of the molecules as they collide with the wall. From a macroscopic point of view, however, we are concerned not with the action of the individual molecules but with the time-averaged force on a given area, which can be measured by a pressure gauge.

1.2 Solids, liquids, and gases

A pure substance is one that has a homogeneous and invariable chemical composition. It may exist in more than one phase, but the chemical composition is the same in all phases. Thus, liquid water, a mixture of liquid water and water vapor (steam), and a mixture of ice and liquid water are all pure substances; every phase has the same chemical composition. In contrast, a mixture of liquid air and gaseous air is not a pure substance.

Consider an amount of ice contained in a weightless piston and cylinder arrangement that maintains a constant atmospheric pressure, as in figure 1.1. Assume that the water starts at $-10\ °C$, in which the state is solid. If ice is slowly heated, the temperature increases and by design in this example the pressure stays constant. When the temperature reaches $0\ °C$ additional heat transfer results in a phase change with the formation of liquid, as indicated in figure 1.1. and the

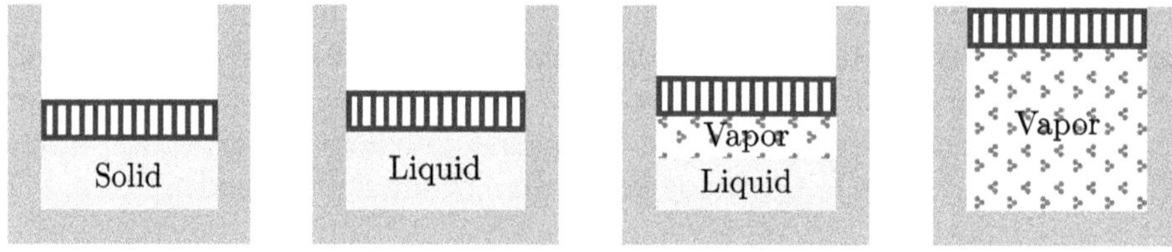

Figure 1.1. Schematic representation of phase changes of a substance. The external pressure exerted on the substance is held constant. When heat is added the solid warms up until its melting point. Further addition of heat melts the solid and the temperature remains constant during melting. After all the solid has melted the addition of heat causes the temperature of the liquid to rise until it reaches the boiling temperature, where the liquid converts to vapor at a constant temperature. After all the liquid is converted to vapor, further addition of heat increases the temperature of the vapor. In the case of water, the density of the solid (ice) is smaller than that of the liquid, therefore the volume of the solid is larger than that of the liquid.

temperature stays constant. Further heating melts all of the solid to a liquid state. If the water is slowly heated, the temperature increases, and the volume increases slightly. When the temperature reaches 99.6 °C additional heat transfer results in a phase change with the formation of some vapor, as indicated in figure 1.1, while the temperature stays constant, and the volume increases considerably. Further heating generates more and more vapor and a large increase in the volume until the last drop of liquid is vaporized. Subsequent heating results in an increase in both the temperature and volume of the vapor, as shown in figure 1.1.

1.3 Important characteristics of substances in various states

1.3.1 The liquid and solid phases

Both liquid and solid states can generally be treated as **incompressible substances** because their specific volume changes very little with temperature. This allows for simpler modeling, as volume is nearly constant across a range of conditions in these phases.

1.3.2 The vapor phase (real gas)

In the vapor phase, molecules experience both attractive and repulsive forces due to neighboring molecules. This molecular interaction makes the behavior of a **real gas** distinct from that of an ideal gas, as intermolecular forces significantly influence its properties.

1.3.3 The ideal gas law

At low densities, gases can often be treated as **ideal gases**, which simplifies the analysis of many thermodynamic processes. The assumptions behind ideal gas behavior are:
1. Molecules do not experience intermolecular forces, meaning they do not attract or repel each other.
2. Molecules are treated as point particles with no volume.

The ideal gas equation, which relates the state variables of an ideal gas, is given by

$$pV = nRT \ \ \text{or} \ \ pV = NkT,$$

where n is the number of moles and N the number of molecules, R is the universal gas constant, and k is Boltzmann's constant.

This equation accurately describes the behavior of gases over a broad range of conditions. However, the ideal gas law significantly differs from equations for liquids and solids (considered incompressible) because an ideal gas's specific volume is highly sensitive to changes in both pressure p and temperature T. Specifically, it varies linearly with temperature and inversely with pressure, reflecting the gas's high compressibility.

The simplicity of the ideal gas law makes it convenient for thermodynamic calculations, although it raises two important questions:

1. What constitutes 'low density' for the ideal gas law to be accurate? In other words, over what density range does this equation of state hold with reasonable precision?
2. How much does a real gas at a specific pressure and temperature deviate from ideal gas behavior?

These questions are essential when assessing whether the ideal gas approximation is valid for a given system.

1.3.4 Internal energy

Internal energy represents the sum of all microscopic forms of energy within a system. It relates to the molecular structure and the level of molecular activity, encompassing the kinetic and potential energies of the molecules.

To better understand internal energy, let us look at a system at the molecular level. In gases, molecules move freely through space, possessing kinetic energy known as **translational energy**. For polyatomic molecules, atoms also rotate around an axis, resulting in **rotational kinetic energy**. Additionally, atoms in polyatomic molecules may vibrate around a common center of mass, which creates **vibrational kinetic energy**. In gases, kinetic energy primarily comes from translational and rotational motion, with vibrational energy becoming significant only at higher temperatures.

On a more detailed level, electrons orbit around an atom's nucleus and possess rotational kinetic energy, with electrons in outer orbits having higher energy. Electrons also spin about their axes, contributing **spin energy**. Other particles, such as neutrons and protons within the nucleus, have similar spin energy. The portion of a system's internal energy associated with the kinetic energies of its molecules is known as **sensible energy**. As temperature increases, molecules move faster and with greater energy, leading to a higher internal energy in the system.

Internal energy also arises from various **binding forces**. These forces exist between molecules within a substance, between atoms within a molecule, and within the particles in an atom's nucleus. Binding forces are strongest in solids and weakest in gases. When a substance absorbs enough energy to overcome these forces, its molecules break free, changing its phase, such as from solid or liquid to gas. This phase-related internal energy is called **latent energy**. Phase changes typically occur without altering the chemical composition of a substance, which is why most practical thermodynamic problems do not require us to consider the forces binding atoms within molecules.

Each molecule contains atoms bound by **chemical energy**, resulting from the bonds between electrons and the nucleus. During chemical reactions, such as combustion, some chemical bonds break while others form, resulting in a change in internal energy. Stronger than these chemical forces are **nuclear forces**, which bind protons and neutrons in the atomic nucleus. The enormous energy associated with nuclear bonds is called **nuclear energy** and only becomes relevant in thermodynamics when considering nuclear reactions, such as fusion or fission. Chemical reactions involve changes in electron structure, but nuclear reactions change the nucleus itself, resulting in the loss of atomic identity.

Atoms may also possess **electric and magnetic dipole-moment energies** when exposed to external electric or magnetic fields. This is due to the twisting of magnetic dipoles caused by the electric currents of orbiting electrons.

Determining the absolute internal energy of a substance (other than an ideal gas) can be complex and challenging, often requiring advanced calculations. Property databases may show large differences in values for internal energy, enthalpy, and entropy (to be discussed later) because these absolute values are somewhat arbitrary. The specific values of enthalpy, entropy, and internal energy at a single state are generally irrelevant; what matters is the difference in these values between two state points. For convenience, many reference handbooks set an arbitrary state point so that the values of these properties remain positive for most common liquid or gas states.

1.3.5 Internal energy of an ideal gas

We can treat a special case to determine the internal energy of an ideal gas, which relies on the concept of **degrees of freedom**. Each degree of freedom represents an independent way in which molecules can store energy. For example, degrees of freedom include the three possible directions of translational motion (movement along the x-, y-, and z-axes), and in multi-atomic gases, additional rotational or vibrational modes.

For **monatomic gases** (such as helium or neon), each molecule has only three translational degrees of freedom because these atoms are treated as single points with no rotational or vibrational motion. According to the equipartition theorem, each degree of freedom contributes $\frac{1}{2}kT$ (where k is Boltzmann's constant and T is temperature) to the energy per molecule. Thus, the internal energy U of N molecules or n moles of a monatomic ideal gas is

$$U = \frac{3}{2}NkT = \frac{3}{2}nRT.$$

For **diatomic gases** (such as oxygen or nitrogen), there are additional degrees of freedom due to rotational motion around two perpendicular axes. Therefore, diatomic gases have five active degrees of freedom at moderate temperatures (three translational and two rotational). The internal energy of a diatomic ideal gas is then

$$U = \frac{5}{2}NkT = \frac{5}{2}nRT.$$

At higher temperatures, vibrational motion also becomes significant, adding more degrees of freedom and increasing the internal energy further. However, at room temperature, this vibrational energy remains negligible for most diatomic gases.

We can generalize the expression for internal energy for an ideal gas as

$$U = f\frac{1}{2}NkT = f\frac{1}{2}nRT = f\frac{1}{2}pV.$$

1.4 System and surroundings

In thermodynamics, a **system** refers to a specific quantity of matter or a region in space selected for study. The **surroundings** include everything outside the system, and the **boundary** is the real or imaginary surface that separates them (see figure 1.2). This boundary can be fixed or movable and acts as the contact surface between the system and its surroundings. The boundary is typically treated as having zero thickness, meaning it contains no mass or volume.

Systems are categorized as closed or open, depending on whether the mass is fixed or if mass and energy can cross its boundaries.

- **Closed system**: A closed system, also known as a control mass, consists of a fixed amount of mass, meaning no mass can enter or leave the system (figure 1.2). However, energy in the form of heat or work can cross the boundary. The volume of a closed system may change. If no energy transfer occurs across the boundary, the system is called an **isolated system**.

  Example: Consider the gas inside a piston–cylinder device shown in figure 1.2. Here, the gas is the system, and the boundary is formed by the inner surfaces of the piston and cylinder. Since no mass enters or exits, it is a closed system, although energy can cross the boundary, and the piston may move. Everything outside the gas, including the piston and cylinder, is considered the surroundings.

- **Open system**: An open system, also known as a **control volume**, is a selected region in space where both mass and energy can cross the boundary. The boundaries of a control volume are referred to as the **control surface**, which can be real or imaginary (figure 1.3).

  Example: If we analyse the flow of air through a tube connecting two cylinders (the right-hand side of figure 1.3), the tube could be selected as a control volume. The inner surface of the tube forms the real boundary, while the entrance and exit areas represent imaginary boundaries. This control volume allows mass and energy to move in and out and enables specific analysis of energy and work interactions, optimizing system components individually.

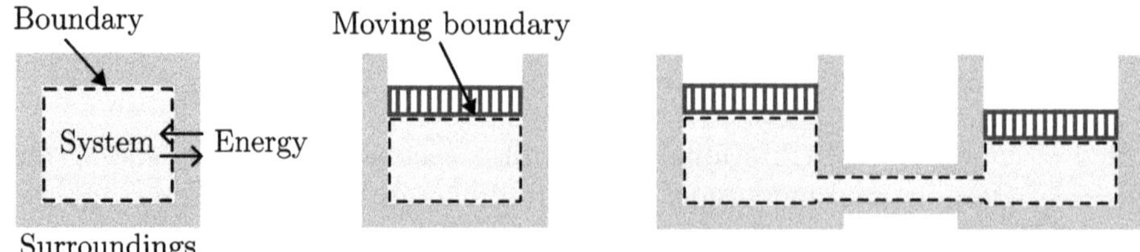

Figure 1.2. Description of a closed system. A thin line (real or imaginary) boundary separates the system from the surroundings. The boundary may move or is firmly fixed. Energy in the form of work or heat can be transferred between the system and the surroundings.

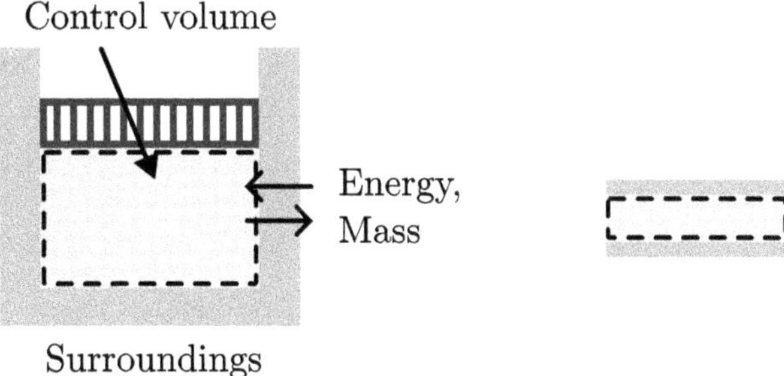

Figure 1.3. Description of a control volume used in the analysis of an open system. Left: Mass can leave the control volume and the control volume may change in shape. In addition to the boundary work due to change in volume of the system, there can be other work, such as shaft or electrical work and heat interaction. Right: A simple tube connecting the two cylinders shown in figure 1.2. By isolating the tube like this, the focus is only on the optimization of this tube. In this case there will be flow in and out and energy interaction with the surroundings. This is the beauty of open system analysis as it allows optimizing individual components of a system.

A control volume may be fixed in size and shape, as with the tube in figure 1.3, or it may involve a moving boundary. Like closed systems, control volumes can experience heat and work interactions, with the added possibility of mass transfer across the boundary.

1.4.1 Examples of closed systems

1. *A sealed water bottle*: When a bottle of water is tightly sealed, it is a closed system. No water or air can enter or leave the bottle, but heat can still transfer across the bottle's surface, affecting the temperature of the water inside.
2. *A pressure cooker*: In operation, a pressure cooker acts as a closed system where no steam or food particles escape (until the pressure relief valve opens). However, heat from the stove transfers through the cooker's walls to raise the temperature inside, cooking the food.
3. *A car battery*: While discharging or charging, a car battery allows electrons to flow through external circuits, transferring energy in the form of electrical work. However, the mass of the battery remains constant, as no material crosses its boundaries, making it a closed system in terms of mass.

1.4.2 Examples of open systems

1. *A boiling pot of water (without a lid)*: When water boils, it allows steam to escape into the air, transferring both mass (water vapor) and energy (heat). The pot is considered an open system because mass and energy cross its boundary.
2. *An air conditioner*: An air conditioner is an open system because it exchanges both mass (air) and energy (cooling effect) with its surroundings. The system intakes warm air, cools it, and releases the cooler air back into the environment.

3. *A car engine*: During operation, fuel and air enter the engine, and exhaust gases are expelled. This makes the engine an open system, as it continuously exchanges mass (fuel, air, and exhaust) and energy (heat and mechanical work) with its surroundings.

1.4.3 Properties of a system

Any characteristic of a system is called a property. Some familiar properties are pressure p, temperature T, volume V, and mass m. The list can be extended to include less familiar ones such as viscosity, thermal conductivity, modulus of elasticity, thermal expansion coefficient, electric resistivity, and even velocity and elevation.

Properties are either intensive or extensive. **Intensive properties** are those that are independent of the mass of a system, such as temperature, pressure, and density. **Extensive properties** are those whose values depend on the size—or extent—of the system. Total mass, total volume, and total momentum are some examples of extensive properties. An easy way to determine whether a property is intensive or extensive is to divide the system into two equal parts with an imaginary partition, as shown in figure 1.4. Each part will have the same value of intensive properties as the original system, but half the value of the extensive properties.

1.4.4 Force balance

Pressure is defined as the normal force exerted by a fluid per unit area. When we discuss pressure, we usually refer to fluids, such as gases or liquids. In solids, the analogous concept is normal stress, although it is important to note that pressure is a scalar quantity, while stress is a tensor. Since pressure measures force per unit area, it is expressed in units of newtons per square meter (N m^{-2}), known as pascals (Pa). This relationship is given by

$$1 \text{ Pa} = 1 \text{ N m}^2.$$

Additionally, 1 bar is equivalent to 10^5 Pa, or 0.1 MPa (100 kPa).

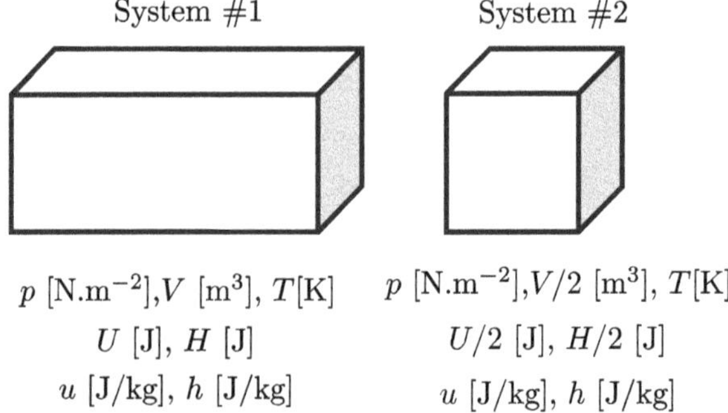

Figure 1.4. The properties of a system can be distinguished as two types: intensive or extensive. Intensive (or effort variables) do not change with the size of the system whereas extensive variables scale proportionally to the size of the system.

1.4.5 Piston–cylinder set-up

Consider the piston–cylinder system shown in figure 1.5. Here, a piston of mass M encloses a gas within a cylinder. To analyse the pressure inside the cylinder, we can apply a force balance on the piston. The forces acting on the piston include:

1. The downward gravitational force due to the piston's weight, Mg,
2. The atmospheric pressure exerted on the piston's surface, $p_{atm}A_p$, and
3. The internal gas pressure acting upwards, pA, where A is the cross-sectional area of the piston.

Setting up the force balance equation,

$$Mg + p_{atm}A - pA = 0.$$

Solving for p, we find that

$$p = p_{atm} + \frac{Mg}{A}.$$

This result shows that the pressure exerted by the gas inside the cylinder depends on both the atmospheric pressure and the weight of the piston. In figure 1.5, the force exerted on the gas is shown as the sum of the atmospheric pressure and the force due to the piston's weight.

1.4.6 Balloon buoyancy and floating conditions

Archimedes' principle states that any object fully or partially submerged in a fluid experiences an upward buoyant force equal to the weight of the fluid it displaces. For balloons, this principle explains how they achieve lift: the balloon displaces a volume of air as it expands. If the weight of the displaced air (the buoyant force) is greater than the combined weight of the balloon and its contents, the balloon will rise. In hot air balloons, this is achieved by heating the air inside, making it less dense. For gas balloons, using a lighter-than-air gas such as helium or hydrogen allows the balloon to displace a greater weight of air than its own weight, creating lift. Archimedes' principle is foundational for understanding how varying densities and volumes allow balloons to float and reach stable altitudes.

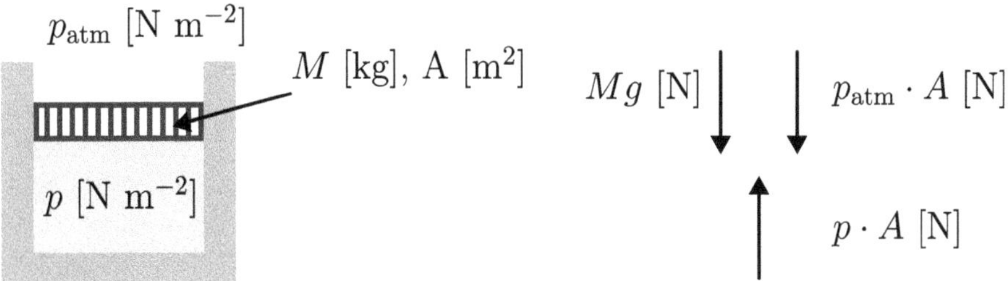

Figure 1.5. Force exerted on the gas is the sum of the force of weight of the piston and the atmosphere.

Let M represent the mass of a balloon at a height z above the ground. This mass includes the gondola, the balloon skin, and any payload, but excludes the gas inside (which could be hot air, hydrogen, or helium). For the balloon to float stably at a height z, the upward buoyant force provided by the displaced air must balance the combined weight of the balloon and the enclosed gas.

This balance condition is given by

$$M + \rho_{in} V = \rho_{air} V,$$

where:
- ρ_{in} is the density of the gas inside the balloon,
- ρ_{air} is the density of the surrounding air, and
- V is the volume of the gas inside the balloon.

If the left side of the equation (representing the total weight) is less than the right side, the balloon will rise; if it is greater, the balloon will descend.

1.4.6.1 Hot air balloons

A hot air balloon operates by heating the air within the balloon, which reduces the density of the air inside compared to the cooler, denser air outside. The balloon is open at the bottom, so the internal pressure equalizes with the external atmospheric pressure. Since warmer air is less dense, we have $T_{in} > T_{air}$, where T_{in} and T_{air} are the internal and external temperatures, respectively, and therefore $\rho_{in} < \rho_{air}$.

1.4.6.2 Gas balloons

Gas balloons, typically filled with hydrogen or helium, operate on a slightly different principle. These gases are naturally less dense than air, allowing the balloon to rise. A gas balloon is initially only partially filled and takes an inverted teardrop shape at ground level. As the balloon ascends, the external air pressure decreases, causing the gas inside to expand. Eventually, the balloon becomes nearly spherical.

If the balloon continues to rise, the internal gas may continue to expand until it reaches the balloon's structural limits, potentially causing it to burst if the balloon material cannot stretch further. For this reason, gas balloons often have mechanisms to release gas or vent it gradually to maintain a stable altitude.

1.4.7 Energy transfer

In classical thermodynamics, energy transfer between a system and its surroundings occurs through two primary modes: heat and work. Consider a system positioned at a certain height above the ground, moving at a constant velocity (figure 1.6). This system, defined by its boundaries, can interact with its surroundings only through these two forms of energy transfer. **Heat transfer** occurs due to a temperature difference between the system and its surroundings, moving energy spontaneously from warmer to cooler regions. **Work transfer**, on the other hand, involves energy transfer due to force applied over a distance, such as when an external force acts on the system or when the system does work on its environment. The energy of the system itself, termed **internal energy**,

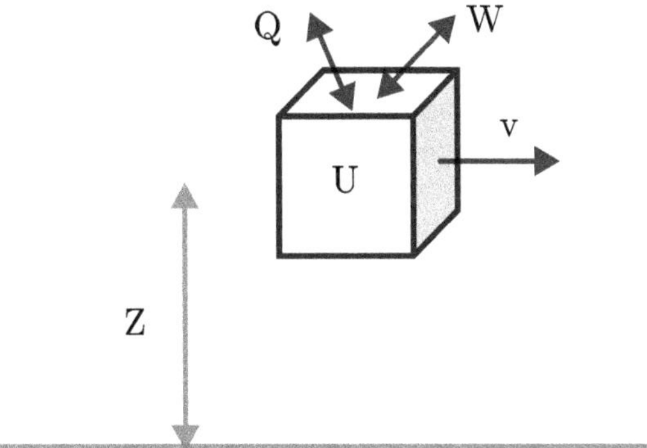

Figure 1.6. Illustration of a system and its interactions with the surroundings. The system, positioned at a height, possesses potential energy due to gravitational forces and kinetic energy due to its velocity. Additionally, it holds internal energy. The system exchanges energy with its surroundings through heat transfer and work transfer.

encapsulates all microscopic kinetic and potential energies within the system and remains distinct from the energy transferred as heat or work. Together, these modes outline the only ways energy can cross the boundary of a system, shaping the fundamental dynamics of thermodynamic processes.

In thermodynamics, work transfer can be broadly categorized into two types: $-pdV$ work and other forms of work transfer. The $-pdV$ work, also known as boundary work, arises specifically from a change in the volume of a system. When a system expands, it performs work on its surroundings, displacing them to create more space. Conversely, when the system compresses or shrinks, the surroundings may do work on the system, exerting force to reduce its volume. Although we often consider the external surroundings as the Earth's atmosphere, they can vary widely; for instance, a system might be in a high-pressure environment or in Martian conditions, where atmospheric pressure is lower than on Earth. On the Moon, with almost no atmosphere, $-pdV$ work would approach zero since there is no significant external pressure. This highlights how $-pdV$ work accounts for work transfer due to volume changes within a system and adapts based on the surrounding pressure, whether atmospheric or otherwise.

In thermodynamics, aside from $-pdV$ work, there are several other types of work transfer modes, each arising from different energy interactions. Here are some of the most common forms:

1. **Shaft work**: Shaft work occurs when a rotating shaft transfers energy into or out of a system, commonly found in engines, turbines, and compressors. For example, in a turbine, the rotating shaft converts fluid energy into mechanical energy, performing work that can drive generators or other machinery. This type of work is calculated based on the torque applied and the angular displacement of the shaft, making it essential in applications where rotational motion is key to energy transfer.

2. **Electrical work**: Electrical work is the transfer of energy due to an electric potential difference. An example would be an electric heater within a system

boundary, where energy is transferred by an electric current to generate heat. The work done is a function of the voltage (potential difference) and charge transfer. This form of work is prevalent in systems where electrical devices operate, such as batteries, electric motors, and circuits.

3. **Magnetic work**: Magnetic work involves the interaction of a system with a magnetic field, where energy is transferred as magnetic field strength aligns or moves magnetic dipoles within a material. An example is a magnetic refrigeration system, where work is done by aligning the magnetic dipoles to cool the system. The work in this case is calculated as the product of magnetic field strength and the magnetic dipole moment.

4. **Dielectric or polarization work**: This type of work occurs when a material's electric dipoles are aligned by an electric field. The interaction between the electric field and the dipole moment can induce work within the system, which is relevant in materials used as capacitors or dielectric insulators. For instance, dielectric heating in microwave ovens uses an alternating electric field to polarize water molecules, resulting in heat production.

5. **Elastic or spring work**: Elastic work, sometimes called spring work, occurs when a force deforms an elastic material, like stretching or compressing a spring. This energy transfer is proportional to the force applied and the displacement, allowing for storage and release of potential energy. It is often seen in applications like suspension systems or mechanical clocks.

Each of these types of work illustrates how diverse energy interactions can facilitate energy transfer across system boundaries.

1.4.8 Moving boundary work

Moving boundary work describes the mechanical work associated with any system that changes its volume, resulting in an energy transfer across its boundaries. This concept applies universally, whether in a piston–cylinder set-up or a more general system where boundaries expand or contract.

Consider a system that, due to internal conditions, begins to expand, increasing its volume. This expansion forces the surroundings to make room for the larger volume, exerting a force on the boundary. As the system volume increases, it performs work on the surroundings by pushing outward against the external pressure. This work transfer, resulting from the volume change, is termed **moving boundary work** and is often expressed as $-pdV$ work, reflecting the product of pressure and the differential change in volume.

Figure 1.7 illustrates two examples of moving boundary work due to volume changes in natural systems:

(a) *A growing tree*: In this scenario, a tree is growing, which means it is gradually increasing in volume. As the tree expands, it displaces the surrounding air, pushing it outward to make space for its growth. This expansion requires the tree (system) to exert a force on the surrounding air (surroundings), effectively performing work on the surroundings. Therefore, in this case, the work transfer occurs from the system (tree) to the surroundings (air).

Figure 1.7. Illustration of boundary work.

(b) *Decaying bananas*: Here, bananas are shown undergoing decay over time, resulting in a gradual reduction in volume. As they decay and shrink, they occupy less space, creating additional volume within the surroundings. Consequently, the surrounding air moves in to fill the space left by the shrinking bananas. In this scenario, the surroundings perform work on the system, as the work transfer is from the surroundings (air) to the system (bananas).

In this book, we adopt a consistent **sign convention for energy transfers**. *Any energy transferred to the system is considered positive*, whether it occurs through heat or work. Conversely, energy transferred from the system to the surroundings is considered negative. This convention simplifies the analysis by clearly indicating whether energy is entering or leaving the system. Since thermodynamics defines only two modes of energy transfer—heat and work—this sign convention applies universally to both. By using this approach, we maintain clarity in interpreting thermodynamic processes and ensure consistency in energy accounting across different scenarios.

In the examples provided, we can use the concept of moving boundary work and our established sign convention to analyse the direction and sign of work transfer.

For the *growing tree* example, our intuition tells us that as the tree expands, it pushes against the surrounding air to make room for its growth. This means that the system (tree) is doing work on the surroundings, transferring energy out of the system. According to our sign convention, this transfer should be negative. The moving boundary work is given by $-pdV$, where p is the atmospheric pressure and dV is the change in volume. In this case, the volume change dV is positive, as the tree's volume increases, while p is constant (atmospheric pressure). Therefore, $-pdV$ yields a negative result, indicating that the work transfer is negative—consistent with our convention for energy leaving the system.

In the *decaying bananas* example, the volume of the system (bananas) decreases over time as they decay, creating additional space that the surrounding air fills. Here, work is done by the surroundings on the system, transferring energy into the system. According to our sign convention, work transferred to the system is positive. In this case, the moving boundary work, $-pdV$, has a negative volume change (dV is negative) since the system's volume is decreasing. Multiplying a negative dV by the negative sign in $-pdV$ results in a positive work value, showing that work is indeed transferred to the system. This result aligns with our sign convention, as energy is entering the system.

These examples illustrate that the $-pdV$ formulation is consistent with our sign convention, accurately reflecting the direction of energy transfer in each case.

1.4.8.1 *Gas enclosed in a cylinder by a piston*

A **special case of work transfer** that resembles moving boundary work occurs when a gas expands within a cylinder enclosed by a piston. In this case, the work transfer is between the gas (system) and the surroundings, specifically through the movement of the piston. Although this work transfer has the same mathematical expression, $-pdV$, it is important to note that this case differs conceptually from the traditional moving boundary work discussed earlier.

In this scenario, p represents the **pressure of the gas inside the cylinder**, not the external atmospheric pressure. As the gas expands, it pushes the piston outward, increasing the system's volume (dV is positive). This expansion requires the gas to perform work on the surroundings (the piston), effectively transferring energy out of the system. According to our sign convention, this work is negative, as energy is leaving the gas. Thus, the expression $-pdV$ accurately represents the direction of work transfer.

Conversely, when the piston compresses the gas, the volume change dV is negative. In this case, the surroundings do work on the gas, transferring energy into the system. The expression $-pdV$ then yields a positive result, indicating that work is being added to the gas, in line with our sign convention.

In summary, while the expression for work transfer $-pdV$ applies here as it does for moving boundary work, *the two scenarios differ*. This case specifically describes work transfer between the gas and its surroundings (the piston), and it relies on the *internal gas pressure* rather than the *external atmospheric pressure*, making it a distinct application of the $-pdV$ relationship.

Building on this specific example, we can calculate the differential work transferred to or from the system by considering the force exerted by the gas pressure on the piston. Consider the gas confined within this piston–cylinder device, as illustrated in figure 1.8. The initial pressure of the gas is p, the total volume is V, and the cross-sectional area of the piston is A. When the piston moves a small distance ds, the differential work W (a scalar quantity) transferred to or from the system during this process is given by

$$W = \vec{F_{\text{ext}}} \cdot \vec{ds} = -\vec{F_{\text{sys}}} \cdot \vec{ds}.$$

To simplify, we replace the force (a vector quantity) and displacement with system properties. The external force exerted on the piston is the opposite of the force exerted by the system, which is equal to the pressure of the gas multiplied by the piston area:

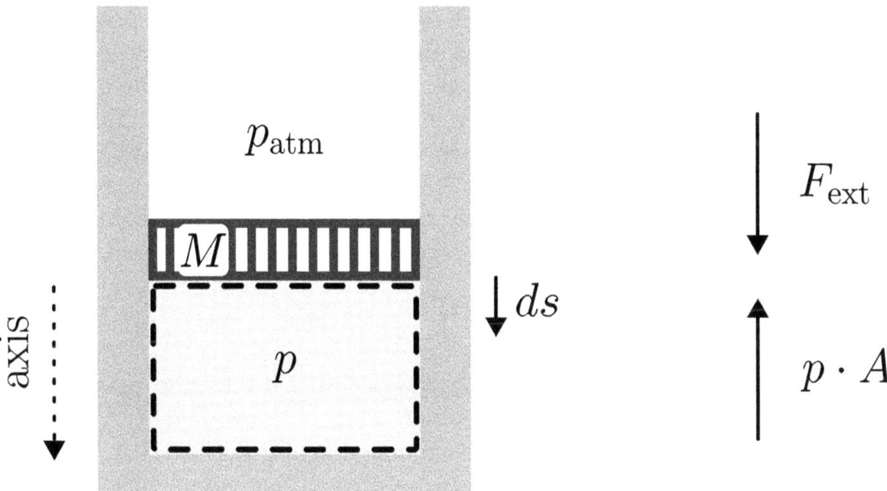

Figure 1.8. Schematic showing the boundary work performed by an external force on the system.

$$\overrightarrow{F_{ext}} = -\overrightarrow{F_{sys}} = -\overrightarrow{pA}.$$

Substituting this into the equation, we obtain

$$W = -\overrightarrow{F_{sys}} \cdot \overrightarrow{ds} = -pA\frac{dV}{A} = -pdV.$$

This transformation elegantly allows us to evaluate the gas work using only system properties (pressure and volume). There is no need to account for the specifics of how the external force is applied. However, it is important to note that the pressure within the system may vary over time. Therefore, to calculate the boundary work over an entire process, we would need to know how the pressure p changes as a function of volume.

It is essential to note that, although the expression $-pdV$ appears the same in both cases, the work calculated for a gas enclosed within a piston–cylinder differs fundamentally from moving boundary work. In traditional moving boundary work, p represents the **external pressure** applied to the system. However, in the specific example of a gas within a piston–cylinder device, p refers to the **internal pressure of the gas** inside the cylinder. This internal gas pressure results from the force exerted by external factors, such as atmospheric pressure and the weight of the piston (as in the above example), but it is an internal property of the system. Thus, while both cases use the $-pdV$ expression, they represent distinct applications, with different interpretations of the pressure term depending on the system's context.

In some books and sources, including online literature, boundary work is defined as pdV without the negative sign. This difference arises from varying sign conventions for energy flows. In this book (and in this course), we adopt the convention that energy flowing *into* the system is positive, while energy flowing *out of* the system is negative. This approach is physically intuitive, as positive energy flow increases the system's energy. Some alternative conventions treat heat transfer to the system as positive but work transfer into the system as negative, with work done by the system (energy transferred out) considered positive. This choice is primarily useful when focusing on the work output of a system but can introduce inconsistency.

The total boundary work over an entire process, as the piston moves from the initial state *i* to the final state *f*, is calculated by integrating all differential work terms:

$$W = -\int_i^f p \, dV.$$

To evaluate this integral, a functional relationship between the pressure p and volume V throughout the process is required.

We show two processes on a pressure–volume diagram in figure 1.9. The process on the left is a constant pressure compression process (the volume decreases). You might wonder how a gas can be compressed at a constant pressure. This is possible by immersing the cylinder in cold surroundings, leading to transfer of heat from the gas to the cold bath. In this diagram, the differential area dA is equal to $p \, dV$. The **differential** work is the *negative* of this area. The total area A under the process curve from an initial state (*i*) to the final state (*f*) is obtained by adding these differential areas.

A gas can follow several different paths as it expands from state *i* to state *f*. In general, each path will have a different area underneath it, and since this area represents the magnitude of the work, the work done will be different for each process. See the process on the right of figure 1.9.

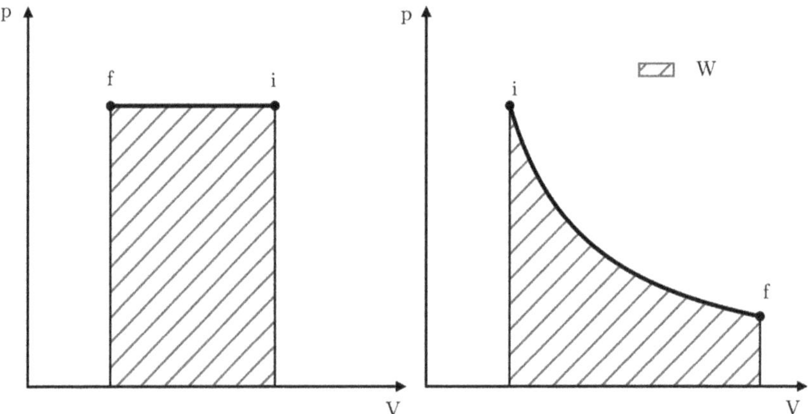

Figure 1.9. A constant pressure (or an isobaric) process (left) and a constant temperature (or an isothermal) process (right) shown on a pressure–volume property diagram. The initial state is indicated by *i* and the final state with *f*. The area under the curve is the negative of the work transferred from the surroundings on the system.

1.5 Summary

In this chapter, we laid the groundwork for classical thermodynamics by exploring fundamental concepts such as energy transfer, system boundaries, and state properties. We examined systems through microscopic and macroscopic lenses, understanding how molecular behaviors aggregate into observable phenomena. Key distinctions between energy in forms such as internal, kinetic, and potential energy were clarified, along with how systems interact with surroundings through heat and work. From the ideal gas law to real gas deviations, we discussed the limitations and applications of these principles in practical scenarios.

IOP Publishing

A Classical Thermodynamics Toolkit

Srinivas Vanapalli

Chapter 2

Enthalpy and heat capacity

Chapter 2 introduces the concept of enthalpy as a foundational component in thermodynamic analysis. Starting with enthalpy, defined as the sum of internal energy and the product of pressure and volume, this chapter highlights its utility in evaluating energy transfer, particularly for open systems where mass and energy flow across boundaries. By establishing enthalpy as an energy potential, this chapter sets the stage for deriving the first law of thermodynamics for both open and closed systems, focusing on how energy in the form of heat, work, and mass flow contributes to a system's energy balance.

Following enthalpy, the chapter delves into heat capacity, exploring how different substances require varying amounts of energy to change temperature. By distinguishing between heat capacity at constant pressure and constant volume, readers will gain tools for analysing how energy interacts with substances under different constraints. This framework prepares readers to tackle more complex thermodynamic applications, particularly those involving energy transfer in flowing systems.

2.1 Enthalpy

By definition enthalpy is

$$H = U + pV,$$

where U represents the internal energy of the system, p is the pressure, and V is the volume. Enthalpy serves as a crucial energy potential, especially useful in the analysis of open systems where mass and energy flow in and out of the system.

To illustrate enthalpy's meaning, let us consider a balloon filled with gas. Initially, the balloon's volume is zero. It is gradually filled with a diatomic ideal gas (with degrees of freedom $f = 5$), reaching a final volume of 1 l, as shown in figure 2.1. The gas pressure and the surrounding atmospheric pressure remain constant at 1 bar. The internal energy difference between the final and initial states is 250 J, and the boundary work done by the gas to expand the balloon is 100 J. (*Note*: this work, as

doi:10.1088/978-0-7503-6029-6ch2   2-1

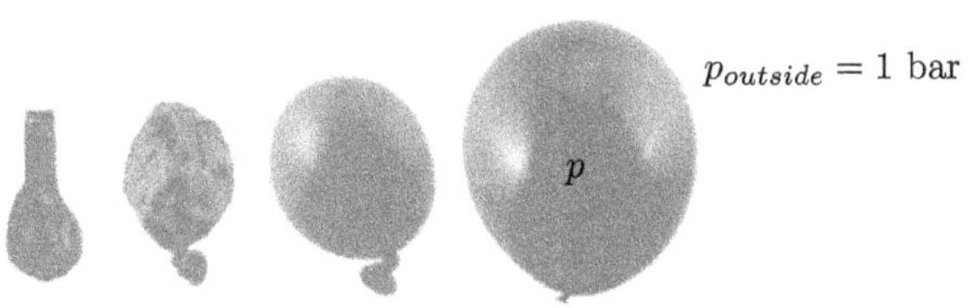

Volume, V	0	1 liter
Internal Energy, U	0	$f\frac{1}{2}pV = 250$ J
Boundary work performed to create the balloon		$W = \int_0^{1liter} pdV = 100$ J
Total energy		$U + W = U + pV = H = 350$ J

Figure 2.1. Example illustrating the physical meaning of enthalpy. See the text for details.

discussed earlier, corresponds to $-pdV$ rather than work done *on* the balloon.) Therefore, the total energy required to inflate the balloon is the sum of the internal energy of the gas and the work done to create the space for it, or $U + pV = 350$ J. This value corresponds to the enthalpy of the ideal gas in its final state.

Therefore, in creating objects 'out of thin air', one must account for both the internal energy and the work done to make space for them. This total energy is the enthalpy. On the surface of the Moon, which lacks an atmosphere, enthalpy is effectively equal to internal energy. On Mars, with its atmospheric pressure of about 6.5 mbar, the enthalpy of gases is lower than it would be on Earth.

2.1.1 Special case: enthalpy of an ideal gas

Using expressions of ideal gas, the enthalpy is

$$H = U + pV = \frac{f}{2}nRT + nRT = \left(\frac{f}{2} + 1\right)nRT \text{ (J)}.$$

$$h_{mol} = \frac{H}{n} = \left(\frac{f}{2} + 1\right)RT \text{ (J mol}^{-1}\text{)}; \quad h_{mass} = \frac{H}{m} = \left(\frac{f}{2} + 1\right)\frac{RT}{M} \text{ (J kg}^{-1}\text{)}.$$

2.1.2 Enthalpy in reference databases

Enthalpy data for most substances is available in property databases. It is important to note that absolute values of enthalpy are not necessary for process analysis. In fact, except for ideal gases, determining absolute enthalpy values is cumbersome and generally unnecessary for system analysis. The REFPROP database, for example, uses the melting point of each substance at one atmosphere as its reference point, deriving all other property data from this baseline. When looking up energy potential values—such as enthalpy, internal energy, Gibbs free energy, or Helmholtz energy—you may notice differences across databases because each may use a distinct reference point.

2.2 Heat capacity of a substance

Experience shows that different substances require different amounts of energy to increase the temperature of identical masses or moles by one degree. This variation leads us to define a property that allows comparison of the energy storage capacities of various substances—**specific heat**. Specific heat is defined as the energy required to raise the temperature of a unit mass or mole of a substance by one degree.

- **Per unit mass:** c_m J (kg K)$^{-1}$.
- **Per unit mole:** c_n J (mol K)$^{-1}$.

The amount of energy needed depends on the process conditions. In thermodynamics, we focus on two main types of specific heats: specific heat at constant volume (c_v) and specific heat at constant pressure (c_p).

- **Constant-volume specific heat (c_v):** This is the energy required to increase the temperature of a unit mass by one degree while keeping volume constant.
- **Constant-pressure specific heat (c_p):** This is the energy needed to raise the temperature of a unit mass by one degree under constant pressure.

Specific heats c_v and c_p can be expressed in terms of energy transferred for any mass or mole of a substance:

energy transferred (J) = mass (kg) $\cdot c_m \cdot \Delta T$ (K) = moles (mol) $\cdot c_n \cdot \Delta T$ (K).

In specific domains, other definitions for specific heat capacity exist, such as specific heat at a constant magnetic field, however, these are outside our current scope.

Now, let us connect specific heats to other thermodynamic properties. Consider a stationary, closed system with a fixed mass undergoing a constant-volume process (meaning no boundary work, $-pdV = 0$). We express the differential energy change per unit mass or mole of the system as:

energy transferred as work or heat per unit mass or mole, $du = c_v dT$ or, equivalently,

$$c_v = \left(\frac{\partial u}{\partial T} \right)_V.$$

Here, u represents the internal energy per unit mass or mole of the substance, maintaining consistent units throughout. Note that while $-pdV = 0$ in a constant-volume process, in addition to heat transfer, energy can still be transferred by other means, such as shaft work or electric work.

For constant-pressure processes, the expression for c_p can be derived similarly. In this scenario, the energy transferred contributes to both the internal energy and the work required to expand the system boundary. Thus, we have:

energy transferred as work or heat per unit mass or mole, $dh = c_p dT$ or, equivalently,

$$c_p = \left(\frac{\partial h}{\partial T}\right)_p.$$

These expressions illustrate that c_v corresponds to changes in internal energy, while c_p relates to changes in enthalpy. To be precise, c_v can be defined as the change in a substance's internal energy per unit temperature change at constant volume, while c_p measures the change in enthalpy per unit temperature change at constant pressure. Thus, c_v provides insight into how internal energy varies with temperature, and c_p similarly reflects enthalpy's temperature dependence.

For an ideal gas, specific heat at constant volume (c_v) and specific heat at constant pressure (c_p) have a straightforward relationship.

Starting from the basic thermodynamic relationship for enthalpy (H),

$$H = U + pV,$$

and knowing from the ideal gas law that

$$pV = nRT,$$

we can substitute $pV = nRT$ into the enthalpy equation, resulting in

$$H = U + nRT.$$

Differentiating with respect to temperature, we obtain

$$dH = dU + nRdT.$$

From definition,

$$nc_p dT = nc_v dT + nRdT,$$

which simplifies to

$$c_p = c_v + R.$$

For an ideal gas, we can further express c_v in terms of the gas's degrees of freedom f. The internal energy U for an ideal gas is

$$dU = nc_v dT = d\left(\frac{f}{2}nRT\right),$$

which gives

$$\frac{f}{2}nRdT = nc_v dT \;\Rightarrow\; c_v = \frac{f}{2}R.$$

For a constant-pressure process, we use the relationship

$$c_p = c_v + R = \frac{f}{2}R + R = \left(\frac{f}{2} + 1\right)R.$$

Thus, c_p for an ideal gas is greater than c_v by R, the gas constant. This difference exists because at constant pressure, energy must also account for the work done in expanding the gas, in addition to increasing its internal energy.

2.2.1 Specific heat of solids and liquids

A substance with a constant specific volume (or density) is considered **incompressible**. For solids and liquids, we typically assume that specific volume or density remains unchanged during a process. This assumption allows us to treat solids and liquids as incompressible substances without significant loss of accuracy. Consequently, for such incompressible substances, the specific heats c_p and c_v can be approximated as equal.

To understand why, let us examine the expression for c_p in terms of enthalpy (h):

$$c_p = \left(\frac{\partial h}{\partial T}\right)_p = \left(\frac{\partial(u + pV)}{\partial T}\right)_p = \left(\frac{\partial u}{\partial T}\right)_p + \left(\frac{\partial pV}{\partial T}\right)_p.$$

Expanding this further, we obtain

$$c_p = \left(\frac{\partial u}{\partial T}\right)_p + p\left(\frac{\partial V}{\partial T}\right)_p + V\left(\frac{\partial p}{\partial T}\right)_p.$$

For incompressible substances, both $\left(\frac{\partial V}{\partial T}\right)_p$ and $\left(\frac{\partial p}{\partial T}\right)_p$ are approximately zero, simplifying the expression to

$$c_p \approx \left(\frac{\partial u}{\partial T}\right)_p \approx c_v.$$

Thus, for solids and liquids, c_p and c_v are effectively the same.

2.2.2 Why is 'heat capacity' not correct?

'Heat capacity' can be confusing and perhaps even misleading, given our modern understanding of thermodynamics. The term originates from a historical context, when 'heat' was not clearly distinguished from other forms of energy transfer. In light of that, 'energy capacity' would arguably be a more accurate term, as it would encompass all forms of energy input, not just heat, that can raise the temperature of a substance.

In fact, reframing 'heat capacity' as 'energy capacity' aligns with the way thermodynamics treats energy transfers in a unified framework. Here's why 'energy capacity' could indeed be a better term:

1. *Inclusive of all energy forms*: Temperature can be increased through various forms of energy transfer, not just heat. For example, both shaft work and microwave radiation can raise a substance's temperature. 'Energy capacity' would recognize that any energy input capable of raising temperature is relevant, whether it is in the form of heat, work, or electromagnetic energy.

2. *Alignment with internal energy*: The concept of internal energy in thermodynamics encompasses both heat and work. Referring to the energy required to change temperature as 'energy capacity' would harmonize this term with the broader understanding of internal energy, rather than implying a special status for heat.
3. *Modern thermodynamics language*: Thermodynamics now distinguishes heat, work, and other forms of energy transfer without giving special emphasis to heat. Using 'energy capacity' could help move away from legacy terms that might reinforce outdated distinctions, making thermodynamics more conceptually accessible.

While tradition maintains 'heat capacity' as a term, a general suggestion to rename it as 'energy capacity' could indeed make thermodynamic concepts clearer and more accurate for learners, aligning with the current understanding that temperature change results from total energy transfer rather than from heat alone.

2.3 Flow work and enthalpy

In open systems, or **control volumes**, mass flows across system boundaries, and work is required to push this mass into or out of the control volume. This work, known as **flow work** (or **flow energy**), is essential to maintain a continuous flow through the control volume.

To derive a relation for flow work, let us consider a small fluid element of volume V (see figure 2.2). The fluid element directly upstream exerts force, pushing this volume into the control volume, much like an imaginary piston would. By assuming the fluid element is sufficiently small, we can treat its properties as uniform.

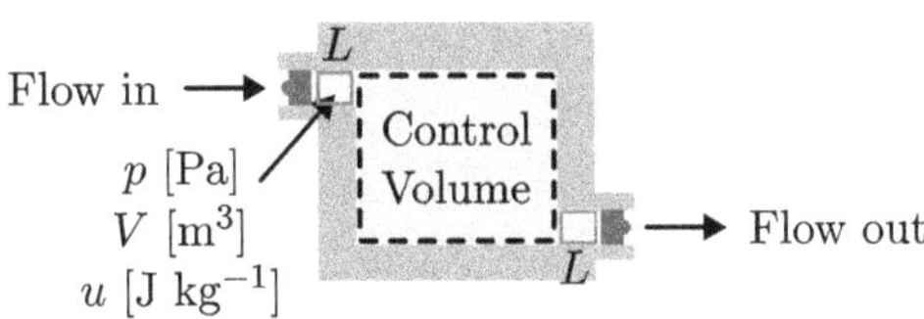

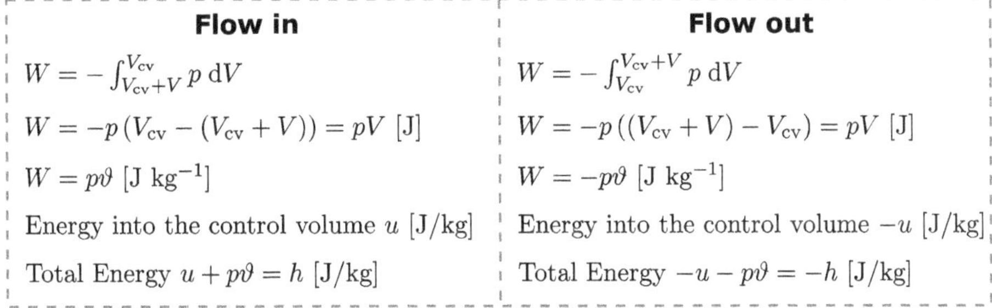

Flow in	Flow out
$W = -\int_{V_{\mathrm{cv}}+V}^{V_{\mathrm{cv}}} p\,\mathrm{d}V$	$W = -\int_{V_{\mathrm{cv}}}^{V_{\mathrm{cv}}+V} p\,\mathrm{d}V$
$W = -p\left(V_{\mathrm{cv}} - (V_{\mathrm{cv}} + V)\right) = pV\ [\mathrm{J}]$	$W = -p\left((V_{\mathrm{cv}} + V) - V_{\mathrm{cv}}\right) = pV\ [\mathrm{J}]$
$W = p\vartheta\ [\mathrm{J\ kg^{-1}}]$	$W = -p\vartheta\ [\mathrm{J\ kg^{-1}}]$
Energy into the control volume $u\ [\mathrm{J/kg}]$	Energy into the control volume $-u\ [\mathrm{J/kg}]$
Total Energy $u + p\vartheta = h\ [\mathrm{J/kg}]$	Total Energy $-u - p\vartheta = -h\ [\mathrm{J/kg}]$

Figure 2.2. Flow work during mass transfer across a control volume boundary. The energy associated with the mass entering or leaving the control volume is represented by the enthalpy of that mass.

If the fluid pressure is p and the fluid element has volume V, the flow work can be expressed as the product pV. It is important to note the sign convention here: work is considered positive when pushing fluid into the control volume and negative when the control volume pushes fluid out. Here, ϑ denotes the specific volume ($\text{m}^3 \text{ kg}^{-1}$).

2.3.1 Total energy of a flowing fluid

The **total energy** of a fluid per unit mass is composed of three primary components,

$$e = u + ke + pe = u + \frac{v^2}{2} + gz,$$

where:
- u is the internal energy,
- $ke = \frac{v^2}{2}$ represents kinetic energy (with v as the velocity), and
- $pe = gz$ is the potential energy (where g is gravitational acceleration and z is the elevation relative to a reference point).

For a flowing fluid, an additional form of energy, **flow energy**, pV, must be included. Thus, the total energy per unit mass for a flowing fluid becomes

$$e_{\text{total}} = u + ke + pe + pv.$$

Since enthalpy h is defined as $h = u + pv$, this relation simplifies to

$$e_{\text{total}} = h + \frac{v^2}{2} + gz.$$

By using enthalpy h instead of internal energy u to represent the energy of a flowing fluid, the flow work pv is automatically included. *This is one of the main reasons enthalpy is defined and commonly used in thermodynamic analysis of open systems.*

2.3.2 Energy transport by mass flow

The total energy associated with a flowing fluid of mass m is given by

$$\text{Total energy transport} = m\left(h + \frac{v^2}{2} + gz\right).$$

For a fluid stream flowing at a mass flow rate $\dot{m}$, the rate of energy transport is

$$\text{Rate of energy transport} = \dot{m}\left(h + \frac{v^2}{2} + gz\right).$$

In many practical situations, the kinetic and potential energy contributions are negligible, allowing these expressions to simplify to mh and $\dot{m}h$ for the total energy transport and rate of energy transport, respectively. This simplification often applies to low-velocity, low-elevation systems where enthalpy alone effectively describes the energy transport.

2.3.3 Examples of flow energy

- *Pumping water into a storage tank*: Flow work is required to pump water up to an elevated storage tank, overcoming both gravitational potential and pressure within the tank.
- *Steam turbines in power plants*: High-pressure steam flows through turbine blades, doing flow work to drive the turbine and generate electricity.
- *Airflow in a building's ventilation system with heat recovery*: Fans push air through ducts, requiring flow work to maintain airflow. Heat recovery units capture thermal energy from outgoing air to warm incoming fresh air, reducing heating needs and increasing efficiency.
- *Compressed air moving through pipes in manufacturing*: Compressed air flows through pipes to power tools and clean surfaces, with flow work enabling continuous supply and pressure maintenance.
- *Fuel flow in a biomass boiler system*: Biomass boilers require fuel to be pushed into the combustion chamber, with flow work ensuring a steady supply of fuel for sustainable heat production.

2.4 Heat transfer definition

To complete our discussion, let us define heat transfer. Heat is the spontaneous flow of energy driven by a temperature difference. There are three main modes of heat transfer—conduction, convection, and radiation—which govern the rate of energy flow:

1. **Conduction**: Transfer of heat through a solid or between solids in contact, where energy moves from molecule to molecule.
2. **Convection**: Transfer of heat by the movement of a fluid, such as air or water, over a surface or within a fluid.
3. **Radiation**: Transfer of heat in the form of electromagnetic waves, which can occur even in a vacuum (for example, the Sun heating the Earth).

Thermodynamics focuses on equilibrium states and establishes the *rules* of energy transfer, while *kinetics*—which we will not cover in this course—determines the *rate* at which energy moves. To illustrate, think of this course as an example: thermodynamics provides the rules for achieving a perfect score, but it is up to you, the student, to study and apply yourself (the kinetics) to achieve that score. How quickly you reach your goal depends on factors such as motivation and focus, just as heat transfer rates depend on the properties of the materials and conditions involved.

2.4.1 First law of thermodynamics

The first law of thermodynamics describes how energy can be transferred to or from a system in three forms: heat, work, and mass flow (figure 2.3). These energy interactions occur at the system boundaries, representing energy gained or lost by the system during a process.

Figure 2.3. The energy of a control volume changes through heat, work, or mass interactions. Multiple inlets or outlets, as well as multiple heat and work interactions, can be present. By convention, energy (heat or work) or mass entering the system is considered positive.

1. **Heat transfer** (Q): Heat transfer to a system (or heat gain) increases the energy of the molecules, raising the internal energy of the system. Conversely, heat transfer from a system (heat loss) decreases the system's energy. Heat transfer occurs when there is a temperature difference, and the energy flows from the higher-temperature area to the lower-temperature area.
2. **Work transfer** (W): Work involves energy interactions that are not caused by a temperature difference between a system and its surroundings. Examples of work include a rising piston, a rotating shaft, or an electric current crossing system boundaries. When work is done on a system (positive work), the system's energy increases. When the system does work on its surroundings (negative work), the system's energy decreases.
3. **Mass flow** (m): Mass flow into and out of a system serves as an additional energy transfer mechanism. When mass enters the system, it brings energy (internal energy) with it, increasing the system's total energy. Conversely, when mass exits the system, it carries energy away, decreasing the system's energy.

2.4.2 Conservation of mass

In control volumes, the **conservation of mass** principle states that the net mass transfer into or out of a control volume over a time interval Δt equals the net change in the total mass within the control volume over Δt:

$$m_{\text{in}} - m_{\text{out}} = \Delta m_{\text{CV}} \ (\text{kg})$$

$$\dot{m}_{\text{in}} - \dot{m}_{\text{out}} = \frac{dm_{\text{CV}}}{dt} \ (\text{kg s}^{-1}).$$

During a **steady-state** process, the total mass in a control volume remains constant (m_{CV} = constant). In this case, the conservation of mass requires that the total mass entering equals the total mass exiting. For a steady-flow system with multiple inlets and outlets, we write this in rate form as

$$\sum_{\text{in}} \dot{m} = \sum_{\text{out}} \dot{m}.$$

This means the total rate of mass entering the control volume is equal to the total rate of mass leaving it.

2.4.3 First law for open systems

For a control volume (open system), the first law of thermodynamics in rate form is practical when dealing with flows. The rate of change of the total energy within a control volume is equal to the net rate of energy entering the control volume via heat, work, and mass flow:

$$\frac{dE_{CV}}{dt} = \sum \dot{Q} + \sum \dot{W} + \sum_{in} \dot{m}\left(h + \frac{V^2}{2} + gz\right) - \sum_{out} \dot{m}\left(h + \frac{V^2}{2} + gz\right).$$

In a **steady-flow** process, the total energy within the control volume remains constant ($E_{CV} = $ constant), so the change in total energy is zero:

$$\sum \dot{Q} + \sum \dot{W} + \sum_{in} \dot{m}\left(h + \frac{V^2}{2} + gz\right) - \sum_{out} \dot{m}\left(h + \frac{V^2}{2} + gz\right) = 0.$$

If changes in kinetic and potential energy are negligible, this simplifies to

$$\sum \dot{Q} + \sum \dot{W} + \sum_{in} \dot{m}h - \sum_{out} \dot{m}h = 0.$$

2.4.4 First law for a closed system

In a closed system, there is no mass flow across the boundaries, so the energy transfer due to mass flow is zero. The first law for a closed system simplifies to

$$\frac{dE_{CM}}{dt} = \sum \dot{Q} + \sum \dot{W}$$

Neglecting kinetic and potential energy changes (e.g. if the system is at rest), the conservation of energy equation reduces to

$$\frac{dU}{dt} = \sum \dot{Q} + \sum \dot{W}.$$

Over a time interval, this can be expressed as

$$\Delta U = \sum Q + \sum W.$$

2.4.5 Examples

1. **Closed system**: heating water in a rigid tank
   - *System description*: A rigid, sealed tank contains a fixed mass of water. The tank is heated from the outside, increasing the water's temperature.
   - *First law application*: Since the tank is sealed, no mass enters or exits (closed system). The first law simplifies to $\Delta U = Q$ because there is no work done (no boundary movement).

- *Relation to $h = c_p T$*: In this case, we primarily use internal energy U rather than enthalpy h, as the volume is constant and there is no mass flow.
- *Key learning*: This example illustrates that in a closed, constant-volume system, the heat added directly changes the internal energy, with no flow or work done by the system.

2. **Open system:** steam flowing through a turbine
   - *System description*: High-pressure, high-temperature steam flows through a turbine, expanding as it drives the turbine blades and exits at a lower pressure and temperature.
   - *First law application*: For this open system with steady flow, the first law in rate form is $\sum \dot{Q} + \sum \dot{W} + \sum_{\text{in}} \dot{m}h - \sum_{\text{out}} \dot{m}h = 0$. In an adiabatic turbine (no heat transfer, $\dot{Q} = 0$) the energy equation becomes $\dot{W} = \dot{m}(h_{\text{out}} - h_{\text{in}})$.
   - *Relation to $h = c_p T$*: For ideal gases, we can approximate enthalpy with $h = c_p T$, allowing us to use temperature data to determine enthalpy change.
   - *Key learning*: This example demonstrates how the first law applies to an open, steady-flow process with energy extracted as work, and how enthalpy changes directly relate to temperature differences.

3. **Closed system:** compressed air in a piston–cylinder assembly
   - *System description*: A piston–cylinder device contains air, initially at low pressure. As the piston compresses the air, its temperature and pressure increase.
   - *First law application*: Here, no mass enters or exits, so we use the closed system form of the first law: $\Delta U = Q + W$. If the process is adiabatic (no heat transfer, $Q = 0$), then $\Delta U = W$, meaning work done on the system increases internal energy.
   - *Key learning*: This example shows how work done on a closed system (in compression) directly affects internal energy and temperature without involving enthalpy.

4. **Open system:** water flowing through a heat exchanger
   - *System description*: Water flows continuously through a heat exchanger, where it is heated from an initial low temperature T_{in} to a higher temperature T_{out}.
   - *First law application*: For a steady-flow open system, the first law becomes $= \dot{m}(h_{\text{out}} - h_{\text{in}})$, assuming no work is done. This shows that heat added equals the increase in enthalpy of the water.
   - *Relation to $h = c_p T$*: With liquids, we can approximate $h \approx c_p T$ if the specific heat c_p is relatively constant over the temperature range.
   - *Key learning*: This example emphasizes how enthalpy change (and thus temperature change) in an open system corresponds to the heat added, ideal for analysing liquid flows in systems such as heating or cooling circuits.

5. **Open system**: compressed air cooling in a nozzle
   - *System description*: Compressed air flows through a nozzle, expanding and cooling as it exits at a higher velocity.
   - *First law application*: In this open system, we apply the first law as $h_{\text{in}} + \frac{v_{\text{in}}^2}{2} = h_{\text{out}} + \frac{v_{\text{out}}^2}{2}$, neglecting heat and potential energy changes. This equation shows that any increase in kinetic energy (velocity) results in a drop in enthalpy, or temperature, of the exiting air.
   - *Relation to $h = c_p T$*: For an ideal gas, we can approximate $h = c_p T$, making it easier to relate temperature drop to changes in velocity.
   - *Key learning*: This example highlights the trade-off between kinetic energy and enthalpy in an open, adiabatic system, illustrating how temperature decreases as velocity increases in expansion processes.

2.5 Summary

In this chapter, we explored enthalpy as an energy potential that is particularly advantageous in analysing open systems. By introducing enthalpy as the sum of internal energy and flow work, we gain a powerful tool for understanding how energy moves in systems with mass exchange. This chapter systematically derived the first law of thermodynamics for open and closed systems, emphasizing the conservation of energy across heat, work, and mass flows.

We also examined flow work in the context of open systems, identifying how flow energy combines with internal energy to create a complete picture of energy movement. The application of heat capacity concepts, particularly at constant pressure and constant volume, helped clarify how energy requirements change with system conditions, especially in the context of ideal gases and incompressible substances.

IOP Publishing

A Classical Thermodynamics Toolkit

Srinivas Vanapalli

Chapter 3

Thermodynamic processes in a closed system

In this chapter, we delve into the core principles of thermodynamic processes within a closed system, exploring how systems transition between states and the energy transformations that accompany these changes. At the heart of this study are **state variables**—key properties such as pressure, temperature, and internal energy—that define a system's state independently of the process path taken. We introduce the concept of **thermodynamic processes** as paths connecting initial and final states, with energy exchanged in the form of **heat** and **work**. This chapter covers various types of processes, including **quasistatic** and **non-quasistatic** processes, as well as specialized cases such as **isobaric**, **isothermal**, **isochoric**, **adiabatic**, and **isentropic** processes. By examining these idealized processes, we build a foundation for understanding real-world applications, with particular emphasis on quasistatic processes involving an ideal gas, where we assume the absence of friction and other forms of non-pV work. This chapter provides essential tools and concepts that will enable students to analyse thermodynamic systems under controlled conditions and understand the implications of idealized versus real behaviors.

3.1 State variables

In thermodynamics, **state variables** (or **state properties**) are essential quantities that describe the current condition of a system. These properties are called 'state' variables because they depend solely on the state of the system at a given moment, not on the process or path taken to reach that state. A defining feature of state variables is that once any two independent state variables are known, other properties of the system can often be derived through established thermodynamic relationships.

For example, in the case of an ideal gas, if the temperature and the amount and type of gas (whether monoatomic or diatomic, for instance) are known, the internal energy U can be determined. Furthermore, if the volume V is also known, the

doi:10.1088/978-0-7503-6029-6ch3

pressure p can be calculated using the ideal gas law, and from these, related properties such as enthalpy H can be deduced.

Common state variables:

- Pressure, p
- Temperature, T
- Volume, V, or density
- Internal energy, U
- Entropy, S
- Enthalpy, H
- Helmholtz free energy, F
- Gibbs free energy, G.

These variables are crucial for analysing the behavior of closed systems under different conditions. Additional state variables may exist, particularly in specialized fields such as electrochemistry or electrical systems, where properties such as electric charge and electric potential are considered.

3.1.1 A note on misconceptions

It is worth mentioning that there is a common misconception where **heat transfer** Q and **work transfer** W are sometimes mistakenly referred to as state properties. This is incorrect, as neither Q nor W are properties of a system; they are forms of energy transfer that depend on the process or path taken during a state change. Unlike state variables, which are path-independent and determined solely by the current state of the system, Q and W cannot be defined by a state alone. Therefore, they are not represented as exact differentials (dQ and dW) but rather as inexact differentials, emphasizing their path-dependent nature. This distinction is crucial for under-standing thermodynamic processes and properly analysing energy transfer within a system.

3.2 Thermodynamic processes

A **thermodynamic process** is defined as a path that connects two states of a system, tracing the sequence of changes that the system undergoes as it moves from an initial state to a final state. During this transition, the system may exchange energy with its surroundings in the form of heat (Q) or work (W). The specifics of how the system changes depend on the conditions imposed, such as temperature, pressure, and volume constraints. Each process follows a unique path on a thermodynamic diagram (e.g. p–V or T–S), indicating how state variables evolve over time. The total energy change in a process depends on both the starting and ending states; however, the way in which energy is transferred—whether through heat, work, or a combination of both—depends on the process path.

It is important to note that thermodynamic processes are not limited to hydraulic or gas systems. In other domains, such as electric or magnetic fields, additional variables such as electric potential and magnetic flux can also define the state and processes of a system, illustrating the broad applicability of thermodynamic principles across various fields of physics and engineering.

3.2.1 Classification of thermodynamic processes

Thermodynamic processes can be classified based on how the system transitions between states, as illustrated in the schematic. Broadly, processes can be categorized as quasistatic or non-quasistatic (figure 3.1). **Quasistatic processes** are those in which the system changes state at a rate much slower than the speed of sound within the system, allowing the system to remain infinitesimally close to equilibrium at each step. This condition ensures that the state variables can be defined consistently throughout the process. These processes can further be divided into reversible and irreversible types. A **reversible process** is an idealized concept where the system and surroundings can be returned to their original states without any net change. **Irreversible processes**, on the other hand, involve inherent dissipative effects such as friction, unrestrained expansion, or heat transfer across a finite temperature difference, making them impossible to reverse without external intervention.

Non-quasistatic processes occur when the system changes at a rate comparable to or faster than the speed of sound, preventing it from maintaining equilibrium during the transition. As a result, significant deviations from a stable state occur throughout the process.

Additionally, while the schematic highlights key classifications, it is important to note that processes can also be categorized as **spontaneous** or **non-spontaneous**. These types are based on the natural tendency of a system to move towards a specific state without external influence. Although spontaneous and non-spontaneous processes are not shown in the current schematic, they are important classifications that will be discussed in greater detail later in the book.

3.2.2 Examples of quasistatic reversible process

1. *Adiabatic expansion of an ideal gas*: A gas confined in a cylinder with a perfectly insulating piston undergoes a slow expansion. The system performs

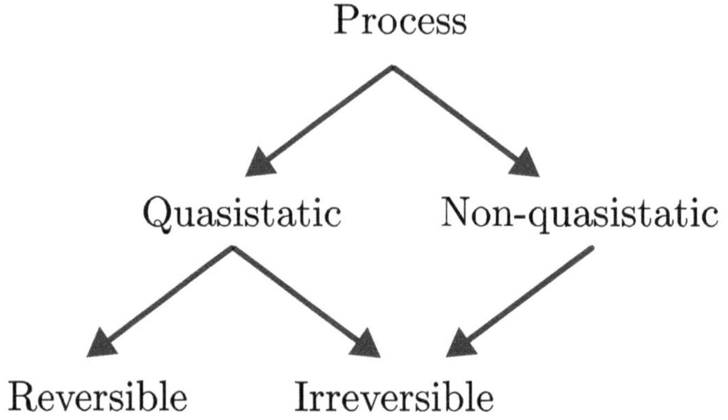

Figure 3.1. Classification of thermodynamic processes.

work on the surroundings without any heat exchange, maintaining a quasistatic and reversible nature due to the controlled, equilibrium-preserving rate of expansion.

2. *Charging of a capacitor*: In an idealized scenario where there is no electrical resistance in the electrical leads, a capacitor connected to a voltage source undergoes a quasistatic, reversible process. The absence of resistance ensures that there is no energy dissipation, allowing the system to stay in a state of electrical equilibrium throughout the charging process.

3. *Compression of a mechanical spring*: An ideal spring being compressed very gradually in a frictionless environment exhibits a quasistatic reversible process. Each incremental compression maintains mechanical equilibrium, and the energy stored in the spring can be fully recovered without loss when the process is reversed.

3.2.3 Examples of quasistatic irreversible processes

1. *Heat transfer across a finite temperature difference*: Consider a system where heat is transferred between a warm body and a cooler body at a controlled, slow rate. Even though the process occurs quasistatically, the finite temperature difference between the system and surroundings means that energy transfer is not perfectly efficient, making the process irreversible.

2. *Isothermal expansion of a gas*: A gas in a cylinder with a piston undergoes an isothermal expansion where it stays at a constant temperature. During this slow, controlled expansion, the system absorbs heat from the surroundings. However, due to the finite temperature difference between the system and the surroundings, the process cannot be perfectly reversed, making it irreversible despite being quasistatic.

3. *Compression of a mechanical lossy spring*: A spring being compressed in a mechanical system with slight internal damping or energy dissipation (such as material hysteresis) represents a quasistatic irreversible process. The slow compression maintains near-equilibrium conditions, but the inherent energy loss within the spring means the process cannot be perfectly reversed.

3.2.4 Examples of non-quasistatic processes

1. *Sudden expansion of gas*: When a gas is released from a high-pressure container into a vacuum or a region of much lower pressure, the process occurs rapidly. This sudden expansion does not allow the system to maintain equilibrium as it changes states, resulting in significant deviations in temperature, pressure, and density throughout the gas. The process is non-quasistatic because the system variables are not uniform during the transition.

2. *Sudden discharge of a capacitor*: When a charged capacitor is suddenly connected across a low-resistance circuit, it discharges almost instantaneously. The rapid flow of current leads to abrupt changes in voltage and current, preventing the system from maintaining equilibrium throughout the

process. The rate of change is so fast that the system's state variables (e.g. electric potential) do not adjust uniformly during the discharge, making it a non-quasistatic process.

3. *Mechanical impact or sudden compression*: When a force is suddenly applied to a mechanical system, such as a hammer hitting a metal surface or a piston being rapidly pushed into a cylinder, the system undergoes abrupt changes in pressure and deformation. The speed of these changes exceeds the rate at which the system can internally equilibrate, resulting in a non-quasistatic process where mechanical and thermodynamic state variables fluctuate unevenly.

3.3 Special cases of processes

In the study of thermodynamics, certain processes are defined by imposing specific constraints on how a system transitions between states. These processes, while idealized, are essential for developing fundamental thermodynamic concepts and analysing the behavior of systems under controlled conditions. Understanding these special cases provides a basis for more complex analyses and helps in recognizing how real systems may approach or deviate from these idealized processes. Below is a summary of common processes and their defining constraints:

- **Isobaric process**: The process occurs at **constant pressure** throughout, often used to study how systems behave when pressure remains unchanged during energy exchanges.
- **Isothermal process**: The process occurs at **constant temperature**, requiring heat transfer to ensure the system temperature does not change despite work being performed.
- **Isochoric process**: The process occurs at **constant volume**, implying that no expansion/compression work is done since the volume of the system does not change.
- **Adiabatic process**: The process involves **no heat transfer** between the system and its surroundings. Energy changes occur solely through work, causing variations in temperature and other state variables.
- **Isentropic process**: The process occurs at **constant entropy**, typically idealized as a reversible adiabatic process where no entropy is generated, allowing for analysis of perfectly efficient systems.

These processes represent idealized models that help establish fundamental thermodynamic principles. However, real systems often exhibit non-ideal behaviors such as friction, heat loss, or imperfect insulation, which can affect the outcomes of these processes. Although we start by studying these idealized processes to build foundational knowledge, later in the book, we will show strategies for analysing real systems that adhere to these constraints but include non-ideal effects. This approach helps bridge the gap between theoretical concepts and practical applications, providing a more complete understanding of how to handle real-world thermodynamic analyses.

3.4 Quasistatic process with a solid

Consider a 100 kg solid with a specific heat capacity $c_s = 0.1$ kJ (kg K)$^{-1}$ initially at a temperature $T_s(t_0) = 500$ K, placed into a liquid reservoir maintained at a constant temperature $T_r = 300$ K. The liquid is assumed to act as an infinite reservoir, implying that its heat capacity is so large that its temperature remains effectively constant throughout the process (figure 3.2).

3.4.1 Example 1: Analysis without considering density change

3.4.1.1 Application of the first law of thermodynamics
To analyse the cooling process of the warm solid when it is immersed in the liquid reservoir, we apply the first law of thermodynamics for a closed system. The first law states that the change in internal energy of a system is equal to the net heat added to the system plus the work done on the system by its surroundings:

$$\Delta U = Q + W.$$

For this initial analysis, we assume there is no significant density change in the solid as it cools. This means any work due to volume change ($W = -p\Delta V$) is negligible, simplifying the equation to

$$\Delta U = Q.$$

3.4.1.2 Analysis of the process
- *Initial condition*:

  ○ The solid has an initial temperature $T_s(t_0) = 500$ K.
  ○ The liquid reservoir, which acts as a constant-temperature heat source, is at $T_r = 300$ K.

- *Assumptions*:

  ○ The process is quasistatic, meaning the heat transfer between the solid and the reservoir is slow enough for the system to remain near thermal equilibrium throughout the process.

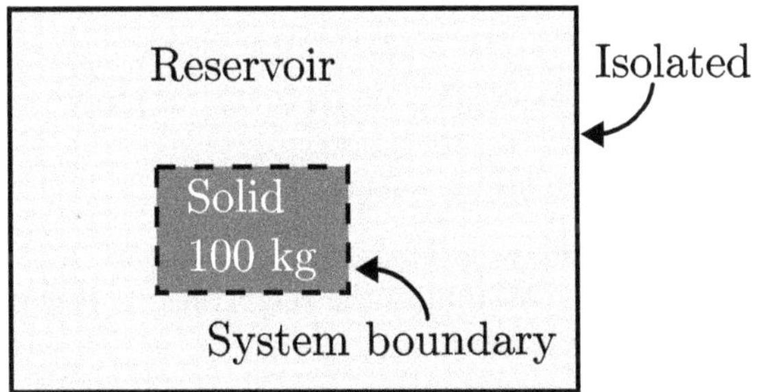

Figure 3.2. Schematic of a hot solid in thermal contact with a reservoir of liquid.

○ The reservoir has a very large heat capacity, so its temperature remains constant at 300 K during the process.
○ Density changes in the solid with temperature are neglected in this analysis.
○ The only form of energy transfer involved is heat exchange (no work interaction).

- *Internal energy change*: The change in internal energy of the solid can be expressed as
  - o $\Delta U = mc_v \Delta T$.
  - o Substituting the given values
  - o $\Delta U = 100 \text{ kg} \times 0.1 \text{ kJ kg K}^{-1} \times (-200 \text{ K}) = -2000 \text{ kJ}$.
- *Heat transfer*: The heat Q transferred out of the solid is equal to the decrease in its internal energy,

$$Q = \Delta U = -2000 \text{ kJ}.$$

The negative value indicates heat flows out of the solid and into the reservoir.

3.4.2 Example 2: Analysis including work due to density change

When the density of the solid changes with temperature, we must consider the work done by the system as it expands or contracts. In this case, the first law of thermodynamics for the solid is

$$\Delta U = Q + W.$$

3.4.2.1 Analysis of the process

- *Initial volume calculation*: To calculate ΔV, we need the initial volume V_0 of the solid. Assuming the density (ρ) of the solid is approximately 8000 kg m^{-3} (a typical value for metals), we can find V_0 as

$$V_0 = \frac{m}{\rho} = \frac{100 \text{ kg}}{8000 \text{ kg m}^{-3}} = 0.0125 \text{ m}^3$$

- *Change in volume*: Using the coefficient of thermal expansion (α),

$$\Delta V = V_0 \alpha \Delta T.$$

Substituting the values,

$$\Delta V = 0.0125 \text{ m}^3 \times 2 \times 10^{-5} \times (-200 \text{ K})$$

$$\Delta V = -0.000\,05 \text{ m}^3.$$

The negative sign indicates a decrease in volume as the solid cools.

- *Work done calculation*: Assuming the reservoir exerts a pressure p of 1 atm (or approximately 101.3 kPa),

$$W = -p\Delta V = -(101.3 \text{ kPa}) \times (-0.000\,05 \text{ m}^3)$$

$$W \approx 0.0051 \text{ kJ}.$$

- *Heat transfer calculation*: Using the first law of thermodynamics,

$$Q = \Delta U - W.$$

Substituting the values,

$$Q = -2000 \text{ kJ} - 0.0051 \text{ kJ}$$

$$Q \approx -2000.0051 \text{ kJ}.$$

The result shows that the heat transferred out of the solid is approximately -2000.0051 kJ. The work term is relatively small compared to the total energy change, but it is included to reflect the slight mechanical energy interaction due to thermal contraction.

Conclusion: In this example, the effect of work due to volume change is minor for the solid, as it undergoes only a small change in volume during cooling. However, including the work term illustrates how the first law of thermodynamics can account for such interactions, providing a more comprehensive analysis. For gases, the work term is typically much larger, as gases expand or contract significantly with temperature changes compared to solids and liquids. This approach becomes especially important for analysing gases or other materials with substantial thermal expansion, as well as for systems under higher pressures where volume changes contribute more prominently to the overall energy exchange.

3.5 Quasistatic process with an ideal gas

In this section, we focus on analysing quasistatic processes involving an ideal gas. These processes are simplified by assuming no additional forms of work, such as electrical or magnetic work, other than pV work. Additionally, we consider idealized conditions where friction is absent, allowing the gas to move smoothly and maintain near-equilibrium at each stage. It is important to note that if heat transfer occurs in any quasistatic process, the process is inherently **irreversible** due to the temperature gradient between the system and its surroundings. The only quasistatic process that remains **reversible** in this context is the **adiabatic process**, where no heat transfer takes place, and the system changes solely through pV work. This is analogous to a **mechanical spring** that compresses and expands without energy loss, provided no other dissipative effects are present. Keeping these assumptions in mind will help clarify the step-by-step analysis presented in this section.

3.6 Isobaric process

In an **isobaric process**, the pressure of an ideal gas remains constant as it undergoes changes in state. Imagine an ideal gas confined in a cylinder with a movable, **frictionless** piston. The pressure on the gas remains constant due to the combined effect of the atmospheric pressure and the weight of the piston. When heat is transferred to the gas, it expands, pushing the piston upward and increasing its volume.

According to the **first law of thermodynamics**,

$$\Delta U = Q + W.$$

In this case, since the gas expands and does work on the surroundings by moving the piston, the work done by the gas can be expressed as

$$W = -p\Delta V,$$

where p is the constant gas pressure, and ΔV is the change in volume of the gas.

Rearranging the first law, we obtain

$$Q = \Delta U - W.$$

3.6.1 Step-by-step analysis

1. *Work done* (W): Since the pressure is constant, the work done by the gas as it expands from an initial volume V_1 to a final volume V_n is

   $$W = -p(V_n - V_1).$$

2. *Internal energy change* (ΔU): For an ideal gas, the change in internal energy depends only on the temperature change and is given by

   $$\Delta U = nc_v\Delta T,$$

   where:
   - n is the number of moles of gas,
   - c_v is the specific heat at constant volume ($\text{J mol}^{-1}\ \text{K}^{-1}$), and
   - $\Delta T = T_n - T_1$ is the temperature difference between the final and initial states.

3. *Substituting in the first law*: Plugging in the expressions for ΔU and W into the first law:

   $$Q = \Delta U - W$$

   $$Q = nc_v\Delta T + p\Delta V.$$

4. *Relating $p\Delta V$ to temperature change*: From the ideal gas law, $pV = nRT$, so we can express $p\Delta V$ as

   $$p\Delta V = nR\Delta T.$$

5. *Final expression for heat transfer*: Substituting $p\Delta V = nR\Delta T$ back into the equation for Q,

   $$Q = nc_v\Delta T + nR\Delta T.$$

Factoring out $n\Delta T$,

$$Q = n(c_v + nR)\Delta T.$$

6. *Defining c_p:* The term $c_p = c_v + R$ is known as the specific heat at constant pressure, so we can write

$$Q = nc_p\Delta T.$$

Thus, the heat transfer in an isobaric process is given by

$$Q = nc_p\Delta T.$$

This formula accounts for both the internal energy change and the work done by the gas as it expands. The specific heat at constant pressure, c_p, includes the extra energy required to perform this work, making it larger than the specific heat at constant volume, c_v.

3.7 Isothermal process

In an **isothermal process**, the temperature of the ideal gas remains constant as the system undergoes changes in state. For an ideal gas, internal energy U depends only on temperature, $U = \frac{f}{2}nRT$. Since the temperature is constant in an isothermal process, there is no change in internal energy, so $\Delta U = 0$.

For an ideal gas at a constant temperature, the first law of thermodynamics is

$$\Delta U = Q + W.$$

Since $\Delta U = 0$ in an isothermal process, this simplifies to

$$Q = -W.$$

This tells us that the heat added to the system is exactly equal (in magnitude) and opposite (in sign) to the work done by the gas.

3.7.1 Step-by-step analysis

1. *Starting with the ideal gas law:* For an ideal gas, the relationship between pressure, volume, and temperature is given by the ideal gas law,

$$pV = nRT.$$

   Since the temperature T is constant in an isothermal process, the product pV also remains constant.

2. *Calculating pV work (W):* As the gas expands or contracts, the pressure and volume change, but the temperature and the product pV remain constant. The work done by the gas when it expands or contracts from an initial volume V_1 to a final volume V_n is given by

$$W = \int_{V_1}^{V_n} -pdV.$$

Using the ideal gas law to express p in terms of V,

$$p = \frac{nRT}{V}.$$

Substituting $p = \frac{nRT}{V}$ into the work integral,

$$W = \int_{V_1}^{V_n} -\frac{nRT}{V}\,dV.$$

3. *Evaluating the integral*: Since n, R, and T are constants in this process, we can pull them out of the integral

$$W = -nRT \int_{V_1}^{V_n} -\frac{1}{V}\,dV.$$

The integral of $\frac{1}{V}$ with respect to V is $\ln V$, so

$$W = -nRT(\ln V_n - \ln V_1).$$

Using the logarithmic identity $\ln a - \ln b = \ln \frac{a}{b}$, we obtain

$$W = -nRT \ln \frac{V_n}{V_1}.$$

Therefore, the pV work done by the gas in an isothermal process is

$$W = -nRT \ln \frac{V_n}{V_1}.$$

4. *Calculating heat transfer Q*: From the first law of thermodynamics, we have

$$Q = \Delta U - W.$$

Since $\Delta U = 0$ in an isothermal process, we find

$$Q = -W.$$

Substituting for W

$$Q = nRT \ln \frac{V_n}{V_1}.$$

5. *Final expressions*: For an isothermal process, the heat transferred Q and the work done W are given by

$$W = -nRT \ln \frac{V_n}{V_1}$$

$$Q = nRT \ln \frac{V_n}{V_1}.$$

The positive sign for Q indicates that heat flows into the gas if it is expanding (since $V_n > V_1$), whereas W is negative, meaning the gas does work on the surroundings.

3.8 Isochoric process

In an **isochoric process**, the volume of the ideal gas remains constant as it undergoes changes in state. Since the volume does not change, the gas does no work on the surroundings (and no work is done on the gas) because $W = -p\Delta V$ and $\Delta V = 0$.

According to the **first law of thermodynamics**:

$$\Delta U = Q + W.$$

In this case, since $W = 0$ in an isochoric process, the first law simplifies to

$$\Delta U = Q.$$

This means that any heat transferred to or from the gas goes entirely into changing its internal energy.

3.8.1 Step-by-step analysis

1. *Work done* (W): Since the volume is constant ($V = V_1 = V_n$), the work done by the gas in an isochoric process is zero:

$$W = \int_{V_1}^{V_n} - pdV = -p.\ 0 = 0.$$

2. *Internal energy change* (ΔU): For an ideal gas, the internal energy U depends only on temperature. The change in internal energy ΔU as the temperature changes from an initial temperature T_1 to a final temperature T_n is

$$\Delta U = U_n - U_1 = \frac{f}{2}nR(T_n - T_1).$$

This can also be expressed using the specific heat at constant volume $c_v = \frac{f}{2}R$,

$$\Delta U = nc_v\Delta T,$$

where $\Delta T = T_n - T_1$ is the change in temperature.

3. *Calculating heat transfer* (Q): Using the first law of thermodynamics, we have

$$Q = \Delta U - W.$$

Since $W = 0$, we obtain

$$Q = \Delta U.$$

Substituting the expression for ΔU,

$$Q = nc_v\Delta T.$$

4. *Final expressions*: In summary, for an isochoric process, the heat transfer Q and the internal energy change ΔU are given by

$$Q = nc_v\Delta T$$

$$\Delta U = nc_v\Delta T.$$

Since the work done $W = 0$, all the energy transferred as heat is used to change the internal energy of the gas.

3.9 Adiabatic process

In an **adiabatic process**, there is no heat transfer between the system and its surroundings, meaning $Q = 0$. Any energy change in the system results entirely from work done on or by the gas.

According to the first law of thermodynamics,

$$\Delta U = Q + W.$$

Since $Q = 0$ in an adiabatic process, this simplifies to

$$\Delta U = W.$$

This implies that the change in internal energy of the gas is equal to the work done by or on the system.

3.9.1 Step-by-step analysis

1. *Starting with the ideal gas law*: For an ideal gas, the relationship between pressure, volume, and temperature is given by the ideal gas law,

$$pV = nRT.$$

2. *Internal energy change (ΔU)*: For an ideal gas, the internal energy U depends only on temperature. The change in internal energy ΔU as the temperature changes from an initial temperature T_1 to a final temperature T_n is

$$\Delta U = nc_v\Delta T,$$

where $c_v = \frac{f}{2}R$ is the specific heat at constant volume, and $\Delta T = T_n - T_1$ is the change in temperature.

3. *Work done (W)*: Since $Q = 0$, all the change in internal energy comes from the work done. Thus

$$W = \Delta U = nc_v\Delta T.$$

4. *Relationship between p and V*: For an adiabatic process in an ideal gas, there is a specific relationship between pressure and volume that can be derived

using the first law. Using the ideal gas law $pV = nRT$ and the relationship $dU = -pdV$, we find that

$$pV^\gamma = \text{constant},$$

where $\gamma = \frac{c_p}{c_v} = \frac{f+2}{f}$ is the heat capacity ratio, often referred to as the adiabatic index.

5. *Calculating pV work*: The work done in an adiabatic process can also be derived using the pV^γ relationship. Integrating W over the process from an initial volume V_1 to a final volume V_n,

$$W = \int_{V_1}^{V_n} -pdV.$$

Using $pV^\gamma = \text{constant}$, we can substitute for p,

$$W = \int_{V_1}^{V_n} -\frac{C}{V^\gamma}dV = -\frac{C}{1-\gamma}(V_n^{1-\gamma} - V_1^{1-\gamma}).$$

Here, $C = p_1 V_1^\gamma = p_n V_n^\gamma$.

6. *Summary of results*: For an adiabatic process:
   - The change in internal energy $\Delta U = nc_v\Delta T$.
   - The work done $W = \Delta U = nc_v\Delta T$.
   - The relationship between pressure and volume is $pV^\gamma = \text{constant}$.

3.10 Arbitrary quasistatic process

An **arbitrary quasistatic process** is one in which an ideal gas undergoes a transition between two states, following no specific constraint on pressure, volume, or temperature. Unlike isothermal, isobaric, isochoric, or adiabatic processes, an arbitrary process does not adhere to a predefined path. Instead, the gas changes state in a general manner, with its pressure and volume varying according to an unspecified functional relationship. By applying the first law of thermodynamics, we can calculate the changes in internal energy, work, and heat transfer for this process, regardless of the exact path taken. This approach allows for a flexible analysis applicable to any quasistatic transition from an initial to a final state, as long as we have sufficient information about the temperature change and the p–V relationship throughout the process.

3.10.1 Step-by-step analysis

1. *Change in internal energy (ΔU)*: For an ideal gas, the internal energy U depends only on the temperature T. The change in internal energy between the initial state (1) and the final state (n) is given by

$$\Delta U = nc_v\Delta T.$$

2. *Work done* (W): In a quasistatic process, the work done by the gas as it expands or contracts from volume V_1 to V_n is given by the area under the curve in a p–V diagram. The pV work is defined as

$$W = \int_{V_1}^{V_n} -p\,dV.$$

To evaluate this integral, we need to know the functional relationship between pressure p and volume V throughout the process. If an explicit expression for p as a function of V is provided (e.g. $p = f(V)$), we can integrate directly. For example:

- **Linear relationship**: If p varies linearly with V, such as $p = aV + b$, we substitute and integrate.
- **Other relationships**: For any other relationship, we follow a similar approach, using the given expression for p in terms of V and integrating over the specified limits.

3. *Heat transfer* (Q): Using the first law of thermodynamics, we can rearrange this to solve for heat transfer Q:

$$Q = \Delta U - W.$$

Substituting the expressions we have for ΔU and W:

$$Q = nc_v\Delta T - \int_{V_1}^{V_n} p\,dV$$

This expression gives the amount of heat transferred during the arbitrary process, accounting for both the change in internal energy and the work done by or on the gas.

3.11 Summary

This chapter provides students with a structured understanding of thermodynamic processes in a closed system, focusing on the following key points:

1. **State variables**: We establish the importance of state variables—properties that depend solely on the system's current state and not on how it reached that state. These variables, such as pressure, temperature, and internal energy, form the basis for analysing changes within a closed system.

2. **Thermodynamic processes**: We define a thermodynamic process as the path taken between two states of a system, highlighting how energy transfer in the form of heat and work varies depending on the process. We examine different types of processes, from simple transitions at constant pressure or temperature to adiabatic and isentropic processes.

3. **Classification of processes**: Processes are categorized into **quasistatic** and **non-quasistatic** types, with quasistatic processes further divided into **reversible** and **irreversible**. We discuss how reversible processes are idealized, with the ability to return both the system and its surroundings to their original states, while irreversible processes involve real-world dissipative effects.

4. **Quasistatic processes with an ideal gas**: Detailed step-by-step analyses are provided for isobaric, isothermal, isochoric, adiabatic, and arbitrary processes, assuming an ideal gas without friction or additional forms of work. We clarify that only the quasistatic adiabatic process can be fully reversible under ideal conditions.

5. **Special process constraints**: We explore the significance of idealized constraints—constant pressure, temperature, volume, and entropy—in defining specific thermodynamic processes, and we caution readers about the limitations of these idealized cases in practical applications.

6. **Energy transfers and the first law**: Using the first law of thermodynamics, we derive expressions for heat transfer, work done, and changes in internal energy for each type of quasistatic process. The chapter emphasizes that in ideal gases, the internal energy change depends solely on temperature, making the analyses applicable across different processes.

By the end of this chapter, students should have a solid grasp of various thermodynamic processes and the tools needed to analyse energy exchanges within closed systems. This foundation prepares them to extend their understanding to the more complex, real-world thermodynamic applications introduced in later chapters.

IOP Publishing

A Classical Thermodynamics Toolkit

Srinivas Vanapalli

Chapter 4

Entropy

In this chapter, we delve into the concept of entropy, a fundamental quantity in thermodynamics that describes the disorder or randomness within a system. We start by exploring the statistical foundation of entropy, viewing it as a measure of possible configurations (microstates) of a system that contribute to an observable state (macrostate). This statistical perspective not only demystifies entropy but also lays the groundwork for understanding it as a state property that only depends on the current condition of a system, independent of its history.

From this foundation, we proceed to define entropy changes under various thermodynamic processes—such as isobaric, isochoric, isothermal, and adiabatic—and examine how each type of process uniquely impacts entropy. By calculating entropy changes in specific, controlled set-ups, we aim to clarify the relationships among heat, work, and entropy in idealized systems. The chapter also discusses arbitrary processes and the conditions required for reversibility and irreversibility, highlighting how entropy dictates the direction and feasibility of natural processes.

Finally, we address the often-misunderstood notion of irreversibility, emphasizing that for a process to be irreversible, the *total* entropy change—accounting for both the system and surroundings—must be positive. This comprehensive view brings us closer to the second law of thermodynamics, which governs the behavior of entropy in all processes.

4.1 Statistical definition of entropy

To understand entropy in statistical terms, let us begin with a simple analogy that introduces the concepts of the microstates and macrostates of a system. Imagine a railway car with three wagons ($N = 3$) transporting a total of three sugar beets ($q = 3$).

doi:10.1088/978-0-7503-6029-6ch4   4-1

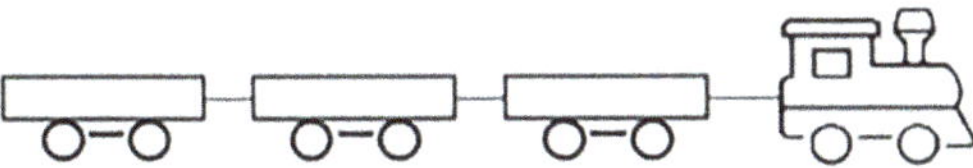 

In this scenario, the beets can be arranged in the three wagons in multiple ways. Each unique arrangement of beets in the wagons is a **microstate** (Ω), and each configuration has an equal probability of occurring. The **macrostate** of the system, on the other hand, is defined by the overall distribution of beets across the wagons— for example, three beets in one wagon, two beets in one wagon and one in another, or one beet in each wagon.

Table 4.1 shows all possible arrangements of the beets among the wagons:
- Three beets in a single wagon—three possible microstates.
- Two beets in one wagon, one in another—six possible microstates.
- One beet in each wagon—one possible microstate.

Thus, we observe three macrostates with different probabilities, as the number of microstates varies for each one. Although each microstate is equally likely, the likelihood of observing a specific macrostate depends on the number of microstates that correspond to it.

For a system with N wagons and q beets, the total number of possible microstates is given by

$$\Omega = \frac{(q + N - 1)!}{q!(N - 1)!}.$$

Table 4.1. All possible arrangements of the beets among the wagons.

	Wagon #1	Wagon #2	Wagon #3
	3	0	0
3 beets in one wagon	0	3	0
	0	0	3
	2	1	0
	2	0	1
	1	2	0
2 beets in one wagon	1	0	2
	0	2	1
	0	1	2
1 beet in one wagon	1	1	1

N_{osc}	q	multiplicity Ω
3	6	28
10	50	6.3×10^{10}
100	1000	10^{144}
10^6	10^7	$10^{1.5 \cdot 10^6}$
$6 \cdot 10^{23}$	10^{26}	$10^{1.6 \cdot 10^{24}}$

Figure 4.1. An example of the astronomical number of microstates in real materials.

Substituting $q = 3$, we calculate $\Omega = 10$, meaning there are ten distinct microstates for this arrangement. Extending this analogy, we can imagine each beet as a unit of energy (or energy quantum) and each wagon as an oscillator within a solid. In real materials, the numbers of oscillators and energy quanta are typically vast, leading to astronomical numbers of microstates, as illustrated in figure 4.1.

4.1.1 Key points from this example

This analogy highlights two crucial aspects of statistical entropy:
1. A system can be described by a large number of equally probable microstates.
2. Observable properties (e.g. counting beets in each wagon) represent macrostates, which differ in likelihood due to varying numbers of corresponding microstates.

4.2 Defining entropy

Entropy (S) is defined as

$$S = k \ln \Omega,$$

where k is the Boltzmann constant, and Ω is the number of microstates. The logarithmic function is used due to the typically immense size of Ω, and the Boltzmann constant serves to scale the units of entropy to joules per kelvin (J K^{-1}).

This statistical view of entropy reveals it as a measure of a system's multiplicity or the number of ways energy can be distributed among particles, reflecting the underlying probability of various macrostates.

4.3 Temperature as a measure of entropy change

Temperature in thermodynamics can be understood through the relationship between **internal energy** and **entropy** when two systems (or bodies) at different initial temperatures come into thermal contact.

Consider two bodies, A and B, each with their own internal energies, U_A and U_B, and entropies, S_A and S_B. When these bodies are brought into thermal contact, energy flows from the body with higher temperature to the one with lower temperature until thermal equilibrium is reached. At equilibrium, no further net energy transfer occurs, and both bodies are at the same temperature.

During this process, the **total entropy** of the combined system, $S_{\text{total}} = S_A + S_B$, changes. Since entropy tends to increase until equilibrium, the process of energy exchange between the two bodies is governed by the maximization of total entropy. The amount of energy exchanged depends on how much each body's entropy changes with a small change in its internal energy.

The change in entropy with respect to internal energy for a body i is given by the derivative $\partial S_i / \partial U_i$. This quantity is fundamental to the definition of temperature. For each body, we define **temperature** as

$$\frac{1}{T_i} = \frac{\partial S_i}{\partial U_i},$$

where:
- T_i is the absolute temperature of body i,
- $\partial S_i / \partial U_i$ is the rate of change of entropy with respect to internal energy.

This relationship means that temperature is a measure of how a body's entropy responds to changes in its internal energy. A system with a high temperature has a relatively small increase in entropy for a given increase in internal energy, while a low-temperature system has a larger entropy change for the same energy addition.

4.4 Interaction and equilibrium

When two bodies A and B interact, energy flows between them in such a way as to increase the total entropy S_{total} until equilibrium is reached. At equilibrium,

$$\frac{\partial S_A}{\partial U_A} = \frac{\partial S_B}{\partial U_B}.$$

By substituting $\frac{1}{T} = \frac{\partial S}{\partial U}$, we find that

$$\frac{1}{T_A} = \frac{1}{T_B}$$

or equivalently, $T_A = T_B$. This condition defines thermal equilibrium: two bodies have the same temperature when they no longer exchange energy and their entropies change in sync with each other's internal energy.

Thus, **temperature** can be defined as the parameter that equalizes the entropy changes per unit of energy between interacting bodies at equilibrium, reflecting a balance in the tendency to disperse energy between them. This statistical view of temperature highlights its role as the factor that regulates entropy exchange in response to energy changes.

4.5 Mechanical equilibrium and the definition of pressure

In thermodynamics, equilibrium is central to defining the measurable properties of a system. In the previous section, we introduced thermal equilibrium as the basis for defining temperature. Now, we extend this concept to mechanical equilibrium, which provides the foundation for defining pressure in a similar way.

4.5.1 Understanding mechanical equilibrium

A system is in **mechanical equilibrium** when no net force acts to change its volume. This means that, under mechanical equilibrium, any infinitesimal change in volume does not lead to a redistribution of energy within the system or between the system and its surroundings. When two systems are in mechanical contact, equilibrium is achieved when they exert equal pressures on each other, preventing spontaneous volume changes.

4.5.2 The definition of pressure

Analogous to the way temperature arises from thermal equilibrium, **pressure** is defined in terms of mechanical equilibrium. Pressure can be understood as a measure of how entropy S changes with volume V at a fixed internal energy U. We define pressure as

$$p = T\left(\frac{\partial S}{\partial V}\right)_U.$$

In this expression:
- p is the pressure of the system,
- T is the temperature, and
- $\left(\frac{\partial S}{\partial V}\right)_U$ represents the rate of entropy change with respect to volume at constant energy.

This relationship shows that pressure is a force per unit area exerted by the system due to its internal molecular motion and interactions. It implies that pressure is a measure of how entropy would increase if the volume were allowed to expand at constant energy.

4.5.3 Interdependence of temperature and pressure

Together, temperature and pressure are both fundamentally related to the system's entropy. Temperature reflects the entropy's sensitivity to changes in energy, while pressure reflects its sensitivity to changes in volume. This unified view of temperature and pressure through entropy provides a powerful framework for understanding equilibrium and thermodynamic processes.

4.6 Entropy is a state property

Entropy, like pressure p, temperature T, volume V, and internal energy U, is a **state property**, meaning it depends only on the current state of the system and not on the specific path taken to reach that state. For an idealized system, entropy S can be expressed as a function of internal energy U, volume V, and particle number N:

$$S = S(U, V, N).$$

This relationship highlights that entropy reflects the possible ways (microstates) in which energy can be distributed within the system at given values of U, V, and N. When any of these variables changes, the number of accessible microstates—and thus the entropy—increases.

4.6.1 Path-independence of entropy changes

Since entropy is a state property, the entropy S of a system depends solely on its current state and not on the process used to reach it. Consequently, the change in entropy ΔS between any two states depends only on the initial and final states, regardless of whether the process was reversible, irreversible, fast, or slow. This **path-independence** is a defining characteristic of state properties in thermodynamics.

Path-independence means that, for any transition between states, the entropy change ΔS remains the same regardless of the process path. Unlike process-dependent quantities such as work and heat, entropy change reflects only the start and end states, which makes it a reliable measure of a system's internal distribution of energy.

This independence from process details is a powerful concept in thermodynamics, making entropy a fundamental property that consistently describes energy dispersal within the system.

4.6.2 Entropy as a function of energy, volume, and particle number

Entropy's dependency on the state variables U, V, and N allows us to understand how changes in these quantities affect entropy.

1. **Energy** (U): Adding energy to a system generally increases entropy, as it allows the particles more possible configurations, increasing the system's **multiplicity**—the number of ways energy can be distributed.
2. **Volume** (V): Increasing the system's volume also raises entropy by allowing more spatial configurations. For example, in a gas, increasing volume gives particles more room to spread out, which expands the number of accessible microstates.

4.6.3 Example: Phases of water—ice, liquid, and vapor

Consider the entropy of 1 kg of water in different phases: ice, liquid, and vapor. For each phase, both the energy U and volume V differ, which directly affects entropy.

- **Ice** has the lowest entropy, as water molecules are tightly organized in a structured lattice, with few accessible microstates.

- **Liquid water** has a higher entropy than ice, as the molecular structure breaks down, allowing for more movement and more possible configurations.
- **Water vapor** has the highest entropy, with molecules moving freely in a large volume, maximizing the multiplicity.

4.6.4 Anomaly of water near freezing point

Water exhibits a unique behavior near 0 °C, where it expands slightly upon freezing. This expansion, caused by hydrogen bonding, increases the volume and slightly increases entropy, even though temperature is decreasing. This anomaly, due to the density difference between ice and liquid water, is crucial for environmental systems.

4.7 Energy and entropy

A fundamental distinction in thermodynamics is that **energy is conserved**, while **entropy is not**. This means that in any process, the total energy change of the Universe—the sum of energy changes in both the system and its surroundings—is always zero, adhering to the **first law of thermodynamics**. Energy can be transferred or transformed, but it cannot be created or destroyed. In contrast, entropy does not follow a conservation law. The **second law of thermodynamics** tells us that the total entropy change of the universe in any process is either zero or positive. This is because, for natural processes, there is an inherent tendency toward increased disorder or greater energy dispersal.

When a process is **reversible**, the entropy change of the Universe is zero; the system and surroundings can return to their initial states without any net change in entropy. However, in **irreversible** processes—which are more common in real-world conditions—entropy increases as energy spontaneously disperses and becomes less available for work. This distinction is crucial: energy is a conserved quantity, strictly balanced in all interactions, while entropy reflects the direction and feasibility of processes, increasing whenever the energy distribution becomes more randomized or less organized. Understanding this difference is essential for grasping why some processes are irreversible and for appreciating the role of entropy in defining the arrow of time.

4.8 Second law of thermodynamics (entropy change in a process)

The second law of thermodynamics describes the change in entropy of a substance during a process. It states that the change in entropy is equal to the sum of the ratio of heat transfer to the system's temperature and the irreversible entropy generated within the system. Mathematically, this is expressed as

$$\Delta S = \int \frac{Q}{T} + S_{\text{irr}},$$

where Q is the heat transfer, T is the temperature of the system, and S_{irr} represents the entropy generated due to irreversibilities in the process.

In differential form, this becomes

$$dS = \frac{Q}{T} + S_{\text{irr}}.$$

In this differential notation, Q represents the infinitesimal heat transfer at each stage of the process. The second law of thermodynamics is a powerful tool for analysing processes, setting thermodynamic limits on various energy conversion processes that we will encounter later in this book.

When the **irreversible entropy** S_{irr} is zero, meaning the process is reversible, we can calculate the entropy change of a substance by using the heat capacity, $c(T)$. For a substance with mass m, the entropy change ΔS is given by

$$\Delta S = \int \frac{Q}{T} = \int \frac{mc(T)dT}{T} = m \int \frac{c(T)dT}{T}.$$

Here, Q represents the heat transferred to the substance, and T is the temperature at each point in the process.

The **heat capacity** $c(T)$ varies depending on the conditions under which heat is added. At **constant volume**, the heat capacity c_v describes how much heat is required to raise the temperature of the substance without changing its volume. Conversely, at **constant pressure**, the heat capacity c_p defines the heat needed to raise the temperature without altering pressure. For ideal gases, c_p is typically greater than c_v because some energy goes into doing work against the surroundings as the gas expands.

This relationship allows us to compute the entropy change by integrating the heat capacity over the temperature range of interest, provided the process is reversible. It is a useful approach for processes where the temperature-dependent behavior of heat capacity, $c(T)$, is known.

4.9 Cooling of a warm solid in a liquid reservoir

Consider a solid with a mass $m_s = 100$ kg and a specific heat capacity $c = 0.1$ kJ kg^{-1} K^{-1} initially at a temperature $T_s(t_0) = 500$ K. This solid is immersed in a large liquid reservoir at a constant temperature $T_r = 300$ K. We assume that the liquid reservoir has an extremely large heat capacity, allowing us to treat its temperature as constant throughout the process.

In this scenario, the solid will transfer heat to the reservoir as it cools down to the reservoir's temperature T_r. This process, where energy flows from a higher temperature to a lower temperature, is inherently irreversible.

Step 1. Heat transfer from the solid to the reservoir

The total heat Q_s transferred from the solid to the reservoir as it cools from $T_s(t_0)$ to T_r can be calculated by

$$Q_s = m_s c(T_r - T_s(t_0)).$$

Substituting the values,

$$Q_s = 100 \text{ kg} \times 0.1 \text{ kJ kg}^{-1} \text{ K}^{-1} \times (300 \text{ K} - 500 \text{ K})$$

$$Q_s = -2000 \text{ kJ}.$$

The negative sign indicates that heat is leaving the solid.

Step 2. Entropy change of the solid (ΔS_s)

The entropy change of the solid as it cools from $T_s(t_0)$ to T_r can be calculated by integrating the infinitesimal entropy changes $dS = \frac{Q}{T}$:

$$\Delta S_s = \int_{T_s(t_0)}^{T_r} \frac{m_s c_s dT}{T} = m_s c_s \ln \frac{T_r}{T_s(t_0)}.$$

Substituting in the values,

$$\Delta S_s = 100 \times 0.1 \times \ln \frac{300}{500}$$

$$\Delta S_s = -5.108 \text{ kJ K}^{-1}.$$

The negative entropy change reflects the decrease in entropy of the solid as it loses heat.

Step 3. Entropy change of the reservoir (ΔS_r)

Since the liquid reservoir's temperature remains constant at $T_r = 300$ K, its entropy change can be calculated using

$$\Delta S_r = \frac{Q_r}{T_r} = \frac{-Q_s}{T_r}.$$

Substituting in $Q_s = -2000$ kJ and $T_r = 300$ K,

$$\Delta S_r = \frac{2000}{300} = 6.667 \text{ kJ K}^{-1}.$$

The positive sign indicates an increase in entropy of the reservoir as it absorbs heat from the solid.

Step 4. Total entropy change (ΔS_{total})

The total entropy change of the universe for this process is the sum of the entropy changes of the solid and the reservoir:

$$\Delta S_{\text{total}} = \Delta S_s + \Delta S_r.$$

Substituting the values,

$$\Delta S_{\text{total}} = -5.108 + 6.667 = 1.559 \text{ kJ K}^{-1}.$$

Since ΔS_{total} is positive, the process is irreversible. This increase in total entropy confirms that the spontaneous flow of heat from the warmer solid to the cooler reservoir cannot naturally reverse.

Key insights:

1. **Conservation of energy**: While energy is conserved (the heat lost by the solid equals the heat gained by the reservoir), entropy behaves differently. The total entropy of the universe increases in this process, indicating irreversibility.

2. **Irreversibility**: The positive entropy change for the universe signifies that this is an irreversible process, as entropy has been generated in the transfer of heat from a higher to a lower temperature.

3. **Temperature gradient and irreversibility**: The irreversibility arises due to the finite temperature difference between the solid and the reservoir. If the solid and the reservoir had the same temperature initially, there would be no heat transfer, and the entropy change would be zero.

This example illustrates a crucial concept in thermodynamics: although energy is conserved in every process, entropy is not. The second law of thermodynamics states that in an irreversible process, the total entropy of the universe always increases, pointing to the fundamental direction of spontaneous processes.

4.10 Entropy production due to heat diffusion in a bar

In this example (figure 4.2), we have a solid bar connecting two thermal reservoirs at different temperatures, with one end at a high temperature $T_H = 400$ K and the other end at a low temperature $T_c = 300$ K. Heat flows through the bar from the hot reservoir to the cold reservoir, maintaining a constant temperature difference. We will calculate the total entropy generation during this process, including the entropy changes of the system (the bar) and the surroundings (the reservoirs).

Assumptions:
1. *Steady-state heat transfer*: The bar operates at steady state, meaning the heat transfer rate $\dot{Q}$ through the bar is constant.
2. *No change in system entropy*: The entropy of the bar itself does not change over time because it is in a steady state, with energy entering and leaving at a constant rate and no accumulation within the bar.

4.10.1 Step-by-step analysis

Step 1. Calculate the rate of heat transfer $\dot{Q}$ using Fourier's law

From Fourier's law of heat conduction, the rate of heat transfer $\dot{Q}$ through the bar is given by

$$\dot{Q} = -\frac{kA}{L}\Delta T,$$

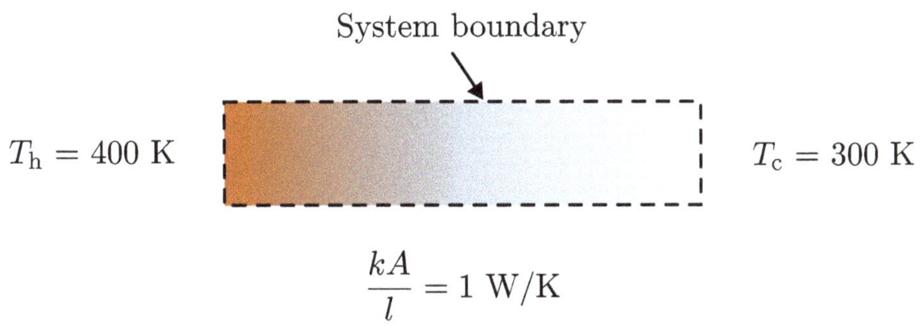

Figure 4.2. A solid bar connecting two thermal reservoirs at different temperatures.

where:
- k is the thermal conductivity of the bar,
- A is the cross-sectional area,
- L is the length of the bar, and
- $\Delta T = T_C - T_H$ is the temperature difference between the ends of the bar.

Given that $\frac{kA}{L} = 1$ (a simplified unit for this problem) and $\Delta T = -100$ K (since heat flows from hot to cold), we find

$$\dot{Q} = -1 \times (-100) = 100 \text{ W}.$$

This means that heat is transferred at a constant rate of 100 W from the hot reservoir to the cold reservoir.

Step 2. Entropy change of the system (bar)

Since the bar is in steady state, there is no net change in its entropy. Heat enters at one end and leaves at the other end, resulting in no entropy accumulation within the bar. Therefore, the entropy change of the system (the bar) is

$$\Delta S_{\text{bar}} = 0.$$

Step 3. Entropy change of the hot reservoir (ΔS_H)

The hot reservoir loses heat at a rate of $\dot{Q} = 100$ W. Since the temperature of the hot reservoir remains constant at $T_H = 400$ K, the entropy change of the hot reservoir per second is

$$\Delta S_H = \frac{\dot{Q}}{T_H}.$$

Substituting the values

$$\Delta S_H = -\frac{100}{400} = -0.25 \text{ W K}^{-1}.$$

The negative sign indicates a decrease in entropy in the hot reservoir as it loses energy.

Step 4. Entropy change of the cold reservoir (ΔS_c)

The cold reservoir gains heat at the same rate, $\dot{Q} = 100$ W, and its temperature is constant at $T_C = 300$ K. The entropy change of the cold reservoir per second is

$$\Delta S_C = \frac{\dot{Q}}{T_C}.$$

Substituting the values,

$$\Delta S_C = \frac{100}{300} = 0.333 \text{ W K}^{-1}.$$

The positive sign indicates an increase in entropy in the cold reservoir as it absorbs energy.

Step 5. Total entropy generation (ΔS_{total})

The total entropy generation in the process is the sum of the entropy changes of the hot and cold reservoirs:

$$\Delta S_{\text{total}} = \Delta S_H + \Delta S_C.$$

Substituting the values from steps 3 and 4:

$$\Delta S_{\text{total}} = -0.25 + 0.333 = 0.083 \text{ W K}^{-1}.$$

Since ΔS_{total} is positive, this confirms that the process is irreversible due to the finite temperature difference between the two reservoirs.

4.10.2 Summary

- **System entropy change (bar)**: The entropy change of the bar is zero because it is in steady state with energy entering and leaving at the same rate.
- **Surroundings entropy change**: The hot reservoir loses entropy, while the cold reservoir gains entropy. The net entropy change of the surroundings (reservoirs) is positive.
- **Irreversibility and entropy generation**: The total entropy generation rate of 0.083 W K^{-1} reflects the irreversibility of heat transfer across a finite temperature difference, consistent with the second law of thermodynamics.

This example illustrates that while energy transfer occurs at a constant rate, entropy is not conserved; it increases due to the irreversible nature of heat conduction through a temperature gradient.

While the example above focuses on Fourier heat conduction, the principle is universally applicable to all modes of heat transfer, including convection and radiation. Heat transfer inherently involves moving energy from a region of higher temperature to a region of lower temperature, which always leads to an increase in the total entropy of the Universe—an indication of irreversibility. This irreversibility holds true regardless of the specific mechanism of heat transfer, as it fundamentally relies on a temperature gradient.

It is worth noting that some authors in the literature describe heat transfer across a very small temperature difference as 'reversible'. While a small temperature difference may indeed result in a very small (or negligible) total entropy change, the process is still fundamentally irreversible. Conceptually, calling it reversible is misleading, as it implies that no entropy is generated. This can confuse readers by suggesting that near-equilibrium heat transfer is exempt from the principles of irreversibility. In reality, even the slightest finite temperature gradient causes entropy production, meaning that all heat transfer processes—no matter how close they are to equilibrium—are irreversible.

4.11 Entropy change in a process: ideal gas in a quasistatic process

In this section, we examine entropy changes during thermodynamic processes, specifically focusing on an ideal gas undergoing a quasistatic transformation. This set-up offers a simplified, clear view of how entropy behaves when no additional work or losses are involved. In a quasistatic process—where each step occurs in perfect control—we gain insight into how entropy shifts under these ideal conditions.

By focusing on the example of an ideal gas, we will see how the system's internal state dictates changes in entropy within this specific set-up. Working in this idealized setting allows us to keep calculations straightforward while connecting physical intuition with mathematical expression. We will break down each part of the process, showing how entropy changes in this carefully controlled transformation.

4.11.1 Isobaric process (constant pressure)

In an isobaric process, where pressure remains constant, we can derive a relationship between temperature and entropy. To do this, we start with the second law of thermodynamics in its differential form, expressing the heat transfer in terms of temperature:

$$dS = \frac{Q}{T} = \frac{nc_p dT}{T}.$$

Integrating this expression yields

$$S = \int \frac{nc_p dT}{T} = nc_p \ln T + C,$$

where C is the integration constant. We can now express temperature as a function of entropy,

$$T = Ae^{S/(nc_p)},$$

where A is a constant that incorporates initial conditions.

If we are interested in the entropy difference between any two states, provided the temperatures of these states are known, we use

$$\Delta S = \int_1^n \frac{Q}{T} = \int_1^n \frac{nc_p dT}{T}.$$

Assuming that the heat capacity, c_p, is independent of temperature, this simplifies to

$$\Delta S = nc_p \ln \frac{T_n}{T_1}.$$

This expression gives us the entropy change for a quasistatic isobaric process between states 1 and n.

In the property diagrams in figure 4.3 illustrate the isobaric quasistatic process between states 1 and n: a pressure–volume (p–V) diagram on the left and a temperature–entropy (T–S) diagram on the right.

4.11.2 Isochoric process (constant volume)

For an isochoric process, where the volume remains constant, we can calculate the entropy change in a similar manner to the isobaric process. Starting with the second

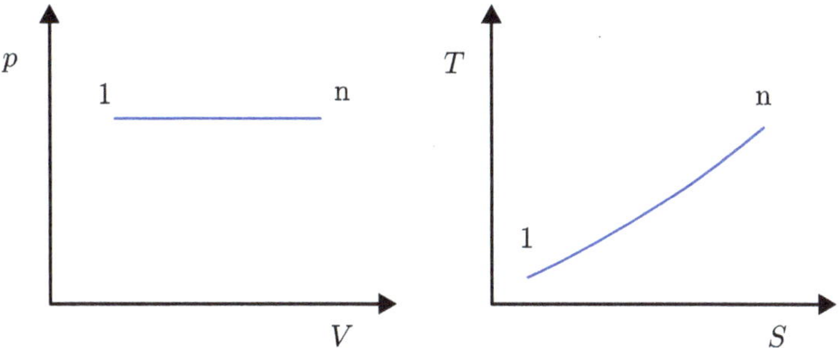

Figure 4.3. Property diagrams for a quasistatic isobaric process between states 1 and n. Left: Pressure–volume (p–V) diagram showing constant pressure. Right: temperature–entropy (T–S) diagram illustrating the entropy increase as temperature rises.

law of thermodynamics in differential form, we express the heat transfer in terms of temperature:

$$dS = \frac{Q}{T} = \frac{nc_v dT}{T}.$$

Integrating this, we find

$$S = nc_v \ln T + C,$$

which allows us to express temperature as a function of entropy,

$$T = Ae^{S/(nc_v)},$$

where A is a constant. For the entropy difference between two states, if temperatures are known and c_V is constant, we have

$$\Delta S = nc_v \ln \frac{T_n}{T_1}.$$

This gives us the entropy change for a quasistatic isochoric process between two states.

4.11.3 Isothermal process (constant temperature)

In an isothermal process, where the temperature remains constant, the change in internal energy is zero. This means that any heat transfer to or from the system is equal to the work done by or on the system, assuming no other types of work are present.

Starting with the second law of thermodynamics in differential form, we have

$$dS = \frac{Q}{T} = \frac{-W}{T} = \frac{-(-pdV)}{T} = \frac{pdV}{T} = \frac{nRdV}{V}.$$

To find the entropy difference between two states with known volumes, we integrate

$$\Delta S = \int_1^n \frac{nRdV}{V}.$$

This simplifies to

$$\Delta S = nR \ln \frac{V_n}{V_1}.$$

Thus, for an isothermal process, the entropy change depends solely on the volume ratio between the two states.

4.11.4 Adiabatic process

In an adiabatic process, there is no heat transfer into or out of the system, and if no other work is done, the entropy change is zero. Since the system is perfectly insulated from its surroundings, no entropy is added or removed, resulting in

$$\Delta S = 0.$$

This means that an adiabatic process is also an isentropic process, where the entropy remains constant throughout the transformation.

4.11.5 Arbitrary process

Entropy, like internal energy, is a state property. This means that the entropy change between two states depends only on the initial and final states, not on the specific path taken. To calculate the entropy difference between any two states, we can follow a similar approach as we did for internal energy by considering a combination of known paths.

In this case, we denote the initial state as 1, the final state as n, and introduce an intermediate state i, as illustrated in figure 4.4.

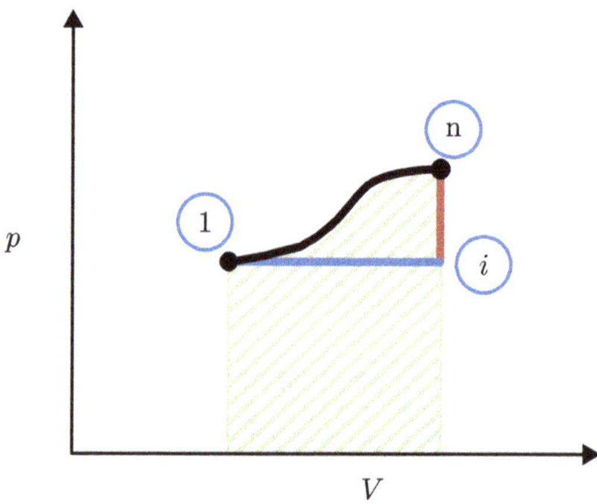

Figure 4.4. Pressure–volume (P–V) diagram of a quasistatic arbitrary process between states 1 and n.

The total entropy change between states 1 and n is the sum of the entropy changes along each segment:

$$\Delta S_{1n} = \Delta S_{1i} + \Delta S_{in}.$$

Using the expressions for entropy change along each segment, we have

$$\Delta S_{1n} = nc_p \ln \frac{T_i}{T_1} + nc_v \ln \frac{T_n}{T_i}.$$

The temperature T_i of the intermediate state, given the pressure p_1 and volume V_n, can be determined using the ideal gas law.

4.12 Condition for an irreversible process

An irreversible process is characterized by a positive *total* entropy change, which includes both the entropy change of the system, ΔS_s, and the entropy change of the surroundings, ΔS_r. This distinction is crucial, as it is sometimes oversimplified in the literature by stating that a process is irreversible when the system's entropy change is positive. However, this is not enough to determine irreversibility. To conclude that a process is irreversible, the *total* entropy change of the combined system and surroundings must be greater than zero:

$$\Delta S_{\text{total}} = \Delta S_s + \Delta S_r > 0.$$

Entropy in a system alone can increase, stay the same, or even decrease during a process, but this does not provide a full picture. We need to consider the surroundings as well, which may also experience an increase, no change, or a decrease in entropy. Only by evaluating the total entropy change of the system and surroundings together can we accurately assess whether a process is truly irreversible. Importantly, the total entropy change can be zero (for a reversible process) or positive, but it can never be negative. This fundamental principle underscores the direction of natural processes and is essential to the correct understanding of entropy and irreversibility.

4.13 Summary

In this chapter, we established entropy as a statistical measure of disorder, introduced its role as a state property, and discussed its calculation across different thermodynamic processes. Key takeaways include:
1. **Statistical definition of entropy**: By connecting microstates and macrostates, we defined entropy as a measure of a system's multiplicity and the probability of various states.
2. **Entropy as a state property**: Like temperature and internal energy, entropy depends solely on a system's state, making its change path-independent.
3. **Entropy change in idealized processes**: We calculated entropy changes in isobaric, isochoric, isothermal, and adiabatic processes, showing how different constraints affect entropy.

4. **Irreversibility and the second law**: We clarified that irreversibility requires a positive total entropy change for the combined system and surroundings, cementing the concept that entropy increases in natural processes.
5. **Scope of idealized systems**: In this chapter, we focused on idealized processes and did not consider non-ideal situations, such as entropy generation due to work dissipation and other irreversible effects.

IOP Publishing

A Classical Thermodynamics Toolkit

Srinivas Vanapalli

Chapter 5

Quasistatic processes with irreversible work transfer and non-quasistatic processes

In this chapter, we delve into the complexities of quasistatic processes with irreversible work transfer and introduce the concept of non-quasistatic processes. Traditional thermodynamic analysis often assumes idealized, reversible conditions, where changes occur slowly enough to maintain equilibrium at each infinitesimal step. However, real-world processes frequently involve irreversible effects—even within quasistatic conditions—that lead to energy dissipation and entropy generation.

We begin by exploring how irreversible work transfer occurs within quasistatic processes, such as frictional and dissipative forces in mechanical systems, and analyse their effects on energy distribution. By separating the reversible and irreversible components of work, we reveal the impact of dissipative forces on internal energy and entropy. To illustrate these effects, we examine a range of ideal gas processes—isochoric, isobaric, isothermal, and adiabatic—that involve irreversibilities, providing insight into entropy generation and the limits of energy recoverability.

The chapter then shifts to non-quasistatic processes, where equilibrium is not maintained, complicating the analysis. Here, classical thermodynamic relationships must be re-examined to account for rapid changes and non-equilibrium conditions. We provide examples, including gas expansion in a vacuum and expansion against constant external pressure, to demonstrate the unique characteristics of non-quasistatic processes and their implications for energy and entropy changes.

Finally, we derive the thermodynamic identity, a fundamental relation that connects internal energy, entropy, temperature, pressure, and volume. This derivation emphasizes a comprehensive approach, respecting the contributions of both heat transfer and irreversibilities, distinguishing it from common simplified treatments. By establishing the thermodynamic identity in this broader context, we reinforce its universal applicability to both reversible and irreversible processes.

doi:10.1088/978-0-7503-6029-6ch5 5-1

5.1 Introduction to quasistatic processes with irreversible work transfer

In thermodynamics, a **quasistatic process** is one that occurs slowly enough for the system to remain in a state of equilibrium at each infinitesimal step. For a process to be **reversible**, however, the total entropy change of the combined system and surroundings must be zero, meaning there is no net entropy generation in the Universe. In a reversible process, both the system and surroundings can, in principle, be returned to their initial states without any net impact on the Universe.

This reversibility is an idealization that assumes the absence of any dissipative effects, such as friction or unrestrained expansion, which would otherwise cause irreversibilities and increase the total entropy of the system and surroundings. In a purely reversible quasistatic process, the entropy of the system may increase, decrease, or remain constant, as long as any changes are exactly balanced by corresponding changes in the surroundings.

Quasistatic, reversible processes are foundational in thermodynamics, serving as a benchmark for ideal energy transformations and allowing us to explore fundamental relationships and limits within thermodynamic systems.

However, in real-world scenarios, processes often contain irreversibilities even when they are slow enough to be considered quasistatic. This introduces the concept of irreversible work transfer within a quasistatic process. Here, the process may still be slow enough to maintain equilibrium at each step, but certain irreversible effects —such as friction, inelastic deformations, or viscous dissipation—prevent the process from being fully reversible. These irreversibilities introduce additional energy losses, which cannot be recovered by simply reversing the process.

Irreversible work transfer has a significant impact on energy distribution within these processes, primarily by causing temperature increases due to dissipative effects such as friction. For example, friction within a piston–cylinder set-up causes mechanical work to raise the temperature of the gas or material, which may lead to heat transfer if the system is not adiabatic. However, in an adiabatic system, this temperature increase would remain contained within the substance without necessarily converting mechanical work directly into thermal energy transferred to the surroundings.

An illustrative example of irreversible work transfer can be seen in a mechanical shaft rotating within a liquid. In this scenario, the shaft's mechanical energy is dissipated as it encounters resistance from the fluid. This interaction increases the temperature of the liquid due to frictional forces, making the work done by the shaft irreversible; it cannot be fully recovered to drive the shaft in reverse. This irreversibility leads to an increase in the total entropy of the system and surroundings, reflecting the energy loss associated with irreversible work transfer.

These examples highlight that even in quasistatic processes, irreversibilities are present, leading to entropy generation and irreversible energy loss within the system and surroundings.

5.2 Energy transfer in quasistatic processes with irreversible work

In a quasistatic process with irreversible work transfer, energy distribution within the system consists of both recoverable and irrecoverable components. While the process is slow enough to assume that equilibrium is maintained at each step, certain irreversible effects—such as friction, resistance, and other non-conservative forces—dissipate a portion of the work. This dissipated work cannot be fully recovered and contributes to entropy generation, setting it apart from purely reversible work in an ideal quasistatic process.

To clarify this, we separate the **reversible** and **irreversible components** of work. The reversible component represents the idealized, recoverable work done by the system that could, in principle, be fully recovered. In contrast, the irreversible component stems from dissipative forces like friction, viscosity, or electrical resistance, resulting in internal energy dissipation and entropy generation within the system.

5.2.1 Examples of reversible and irreversible work

Consider a **battery** as an example of reversible work transfer. In an ideal scenario, the battery converts chemical energy into electrical work, which could be recovered by recharging, making it a reversible process. However, in real batteries, **Joule heating** due to internal resistance and **electrode losses** are irreversible components, as they result in entropy generation and energy loss. These effects cannot be fully reversed, even if the battery is recharged.

Another example is a **fuel cell** that combines hydrogen and oxygen to produce electricity and water. Ideally, this process is reversible, as electricity could be used to split water back into hydrogen and oxygen, allowing complete recovery of the input energy. However, in practice, **Joule heating, electrode degradation**, and **formation of gas bubbles** within the electrolyte lead to irreversible energy losses that cannot be fully recovered, making part of the work irreversible.

A third example is the **expansion of a gas within a piston–cylinder set-up**. In an ideal scenario, the work done by the gas expanding against the piston could be recovered completely if there were no frictional losses. However, with a **lossy piston**, frictional effects dissipate some of the work, which increases the internal energy of the gas or piston material. This irreversibility results in an increase in the total entropy of the system and surroundings, and the dissipated work cannot be recovered.

5.2.2 Mathematical treatment of reversible and irreversible work

Using the first law of thermodynamics as defined in this book, we can write

$$\Delta U = Q + W,$$

where ΔU represents the change in internal energy of the system, Q is the heat added, and W is the total work done on the system.

In a quasistatic process with irreversibilities, we can separate W into the reversible work, W_{rev}, and irreversible work, W_{irr}, components:

$$W = W_{\text{rev}} + W_{\text{irr}}.$$

The **reversible work**, W_{rev}, represents the ideal component of work that could, in principle, be fully recovered. The **irreversible work**, W_{irr}, on the other hand, arises due to dissipative forces such as friction or resistance, leading to energy dissipation within the system.

In the modified energy equation, the presence of W_{irr} reflects the energy lost due to irreversibilities:

$$\Delta U = Q + W_{\text{rev}} + W_{\text{irr}}.$$

In a purely reversible process, W_{irr} would be zero, meaning there would be no net entropy generation and all the work could theoretically be recovered. However, when W_{irr} is non-zero due to dissipative forces, there is an irreversible loss of energy, contributing to an increase in total entropy.

By separating W_{rev} and W_{irr} in the energy equation, we gain insight into how energy transfers are affected by irreversibilities in quasistatic processes. These examples and equations demonstrate the distinct roles of reversible and irreversible work, and underscore the impact of irreversibilities on entropy and energy transfer within thermodynamic systems.

5.3 Entropy generation in quasistatic processes with irreversible work

In thermodynamic processes, **entropy generation** is a direct indicator of irreversibility. Whenever a process involves dissipative effects such as friction, electrical resistance, or inelastic deformation, it becomes irreversible, leading to a net increase in the total entropy of the system and surroundings. This increase in entropy, or entropy generation, is the hallmark of irreversible work within a thermodynamic process, including those that are quasistatic.

5.3.1 Entropy generation due to irreversibilities

In an **idealized reversible process**, the total entropy change of the system and surroundings is zero, meaning no net entropy is generated, and the process could theoretically be reversed without any loss of energy or increase in entropy. However, in a **quasistatic process with irreversible work**, dissipative forces introduce irreversibilities that generate entropy. This entropy generation reflects the portion of the energy that cannot be recovered, as it has been dissipated in ways that increase the randomness or disorder of the system.

5.3.2 Relationship between irreversible work and entropy production

The presence of irreversible work in a quasistatic process directly contributes to entropy generation. For instance, consider a gas expanding in a piston–cylinder arrangement with friction. The work done by the gas against the piston includes an irreversible component due to the frictional forces between the piston and cylinder wall. This irreversible work, W_{irr}, results in increased temperature and disorder within the system, which cannot be reversed without adding energy from an external source. Consequently, this irreversible work leads to a **positive entropy generation** term, representing the loss of available energy in the system.

The relationship between entropy generation S_{gen} and irreversible work W_{irr} can be expressed as

$$S_{gen} \propto W_{irr}/T,$$

where T is the absolute temperature at which the irreversibility occurs. This relation implies that the greater the irreversible work, the more entropy is generated. The entropy generation term, S_{gen}, measures the degree of irreversibility in the process and signifies the portion of the energy transformation that is irrecoverable.

5.3.3 Entropy in reversible versus irreversible quasistatic processes

In a **purely reversible quasistatic process**, there are no dissipative forces, and as a result, there is no entropy generation. Any entropy change within the system is exactly balanced by an equal and opposite change in the surroundings, leading to zero net entropy change in the Universe. This reversibility implies that all work performed can be fully recovered, with no energy lost to irreversible effects.

In contrast, in a **quasistatic process with irreversible work**, entropy is generated within the system or surroundings (or both), leading to a net positive entropy change in the Universe. This irreversibility reduces the potential for energy recovery, as part of the work has been dissipated through mechanisms that increase entropy, such as frictional heating or resistive heating. The resulting entropy generation thus distinguishes quasistatic processes with irreversible work from ideal reversible processes, underscoring the fundamental difference in energy efficiency and recoverability.

In summary, the presence of entropy generation in a quasistatic process with irreversible work signals the loss of usable energy and highlights the impact of irreversibilities. This entropy generation reflects the inherent inefficiencies in real-world processes, contrasting sharply with the idealized behavior of purely reversible systems.

5.4 Quasistatic irreversible ideal gas processes

This section explores various types of ideal gas processes that are quasistatic but involve irreversibilities. In each case, the process is slow enough to maintain equilibrium at every step, yet friction, heat transfer resistance, or other dissipative effects introduce irreversibilities, resulting in entropy generation.

5.4.1 Isochoric process with irreversibility (constant volume)

In an isochoric process, the volume of the gas remains constant, so no boundary work is done ($W = 0$). We will analyse two cases where the system undergoes a change in internal energy due to heat transfer, and in one case, additional work through a rotating shaft. In both cases, we assume the gas is ideal, and the system goes from the same initial state to the same final state.

Let T_{sur} be the temperature of the surroundings, which remains constant during the process.

Case 1: Isochoric process with only heat transfer
In this first case, the gas undergoes an isochoric process where heat Q_1 is transferred from the surroundings to the system. According to the first law of thermodynamics as defined in this book,

$$\Delta U = Q_1,$$

where ΔU is the change in internal energy of the gas. Given that volume is constant, all the heat transfer goes into changing the internal energy.

For an ideal gas undergoing an isochoric process, the entropy change of the system is

$$\Delta S_{\mathrm{system}} = nc_v \ln\left(\frac{T_f}{T_i}\right),$$

where T_i and T_f are the initial and final temperatures of the gas, n is the number of moles, and c_v is the molar heat capacity at constant volume.

Since this process is irreversible (likely due to a temperature gradient between the surroundings and system), there is entropy generation in the surroundings, denoted $S_{\mathrm{gen},1}$. For the surroundings, the entropy change due to heat transfer Q_1 at temperature T_{sur} is

$$\Delta S_{\mathrm{surroundings},1} = -\frac{Q_1}{T_{\mathrm{sur}}}.$$

The total entropy generation for case 1, combining the system and surroundings, is

$$S_{\mathrm{gen},1} = \Delta S_{\mathrm{system}} + \Delta S_{\mathrm{surroundings},1} = nc_v \ln\left(\frac{T_f}{T_i}\right) - \frac{Q_1}{T_{\mathrm{sur}}}.$$

Case 2: Isochoric process with heat transfer and shaft work
In the second scenario, in addition to heat transfer, a shaft rotates within the gas, performing mechanical work W_{shaft} on the gas. This work is dissipated entirely as heat within the system, contributing to the internal energy change and reducing the amount of heat Q_2 needed from the surroundings.

The total change in internal energy is the same as in case 1:

$$\Delta U = Q_2 + W_{\mathrm{shaft}}.$$

Since both cases start and end at the same states, we have

$$Q_1 = Q_2 + W_{\mathrm{shaft}}.$$

Therefore, $Q_2 < Q_1$ due to the additional energy input from the shaft work.

For the system, the entropy change remains the same as in case 1, as it only depends on the initial and final states:

$$\Delta S_{\text{system}} = nc_v \ln\left(\frac{T_f}{T_i}\right).$$

For the surroundings, the entropy change due to the heat transfer Q_2 is

$$\Delta S_{\text{surroundings},2} = -\frac{Q_2}{T_{\text{sur}}}.$$

Comparison of entropy generation

The total entropy generation for case 2, combining the system and surroundings, is

$$S_{\text{gen},2} = \Delta S_{\text{system}} + \Delta S_{\text{surroundings},2} = nc_v \ln\left(\frac{T_f}{T_i}\right) - \frac{Q_2}{T_{\text{sur}}}.$$

Since $Q_2 < Q_1$, we have

$$-\frac{Q_2}{T_{\text{sur}}} > -\frac{Q_1}{T_{\text{sur}}}.$$

Therefore,

$$S_{\text{gen},2} > S_{\text{gen},1}.$$

This shows that the additional shaft work in case 2 introduces more irreversibility, leading to a greater entropy generation than in case 1. Although less heat is needed from the surroundings in case 2, the total irreversibility of the process is higher due to the dissipative work done by the shaft. This example highlights how work dissipation can increase entropy generation, even when reducing external heat transfer.

5.4.2 Isobaric process with irreversibility (constant pressure)

In an isobaric process, the pressure of the gas remains constant, which allows for boundary work as the gas expands or contracts. We will consider two cases where the system undergoes an internal energy change due to heat transfer, and in one case, additional work is provided through a rotating shaft. The gas is assumed to be ideal, and both cases go from the same initial to the same final state.

Let T_{sur} represent the temperature of the surroundings, which remains constant throughout the process.

Case 1: Isobaric process with only heat transfer

In the first scenario, heat Q_1 is transferred to the system from the surroundings at constant pressure, causing the gas to expand. According to the first law of thermodynamics as defined in this book,

$$\Delta U = Q_1 + W,$$

where ΔU is the change in internal energy, Q_1 is the heat transfer, and W is the work done by the system on the surroundings. For an isobaric process, the work done by the gas on the surroundings is given by

$$W = -p\Delta V.$$

Since ΔU depends only on the initial and final states, the heat transfer Q_1 is required to account for both the internal energy change and the work done,

$$Q_1 = \Delta U + pdV.$$

The entropy change of the system, given it is an ideal gas, can be expressed as

$$\Delta S_{\text{system}} = nc_p \ln\left(\frac{T_f}{T_i}\right),$$

where T_i and T_f are the initial and final temperatures, n is the number of moles, and c_p is the molar heat capacity at constant pressure.

Because this process is irreversible (possibly due to a temperature gradient between the surroundings and the system), there is entropy generation in the surroundings, which we denote as $S_{\text{gen,1}}$. For the surroundings, the entropy change due to heat transfer Q_1 at temperature T_{sur} is

$$\Delta S_{\text{surroundings,1}} = -\frac{Q_1}{T_{\text{sur}}}.$$

The total entropy generation for case 1, combining the system and surroundings, is

$$S_{\text{gen,1}} = \Delta S_{\text{system}} + \Delta S_{\text{surroundings,1}} = nc_p \ln\left(\frac{T_f}{T_i}\right) - \frac{Q_1}{T_{\text{sur}}}.$$

Case 2: Isobaric process with heat transfer and shaft work
In the second scenario, in addition to heat transfer, a shaft is rotating within the gas, performing mechanical work W_{shaft} on the gas. This work is fully dissipated as frictional heat within the system, supplementing the heat input required from the surroundings.

The total change in internal energy remains the same:

$$\Delta U = Q_2 + W_{\text{shaft}} + W.$$

Since both cases start and end at the same states, we have

$$Q_1 = Q_2 + W_{\text{shaft}}.$$

Thus, in this case, the heat transfer Q_2 from the surroundings is lower than Q_1 as part of the energy input is provided by the shaft work W_{shaft}.

The entropy change of the system, given it is an ideal gas and depends only on the initial and final states, remains the same as in case 1:

$$\Delta S_{\text{system}} = nc_p \ln\left(\frac{T_f}{T_i}\right).$$

The entropy change of the surroundings, due to the heat transfer Q_2, is

$$\Delta S_{\text{surroundings},2} = -\frac{Q_2}{T_{\text{sur}}}.$$

Comparison of entropy generation
The total entropy generation for case 2, combining the system and surroundings, is

$$S_{\text{gen},2} = \Delta S_{\text{system}} + \Delta S_{\text{surroundings},2} = nc_p \ln\left(\frac{T_f}{T_i}\right) - \frac{Q_2}{T_{\text{sur}}}.$$

Since $Q_2 < Q_1$, we have

$$-\frac{Q_2}{T_{\text{sur}}} > -\frac{Q_1}{T_{\text{sur}}}.$$

Therefore,

$$S_{\text{gen},2} > S_{\text{gen},1}.$$

This analysis shows that the additional shaft work in case 2 increases the total entropy generation. While less heat is required from the surroundings, the irreversibility associated with the dissipative shaft work results in greater entropy generation, highlighting the impact of energy dissipation on entropy production in an isobaric process.

5.4.3 Isothermal process with irreversibility (constant temperature)

In an isothermal process, the temperature of the gas remains constant, so the internal energy change is zero for an ideal gas ($\Delta U = 0$). For this analysis, we assume that the surroundings are at a slightly higher temperature $T_{\text{sur}} = T + \Delta T$ than the system's temperature T. This slight temperature difference drives heat transfer from the surroundings to the system, allowing the gas to perform work while maintaining an isothermal state.

Case 1: Isothermal process with only heat transfer
In the first case, the gas undergoes an isothermal expansion where the work done on the surroundings is entirely supplied by heat transfer Q_1 from the surroundings. According to the first law of thermodynamics,

$$\Delta U = Q_1 + W.$$

Since $\Delta U = 0$ for an isothermal process, we have

$$Q_1 = -W.$$

The work done by the system during an isothermal expansion from an initial volume V_i to a final volume V_f at temperature T is given by

$$W = -nRT \ln(V_f/V_i).$$

Therefore, the heat transfer required is

$$Q_1 = nRT \ln(V_f/V_i).$$

The entropy change for the system, due to the isothermal heat transfer, is

$$\Delta S_{\text{system}} = \frac{Q_1}{T} = nR \ln(V_f/V_i).$$

Since this process is irreversible due to the temperature difference ΔT between the system and surroundings, entropy is generated. For the surroundings, the entropy change due to the heat transfer Q_1 at temperature T_{sur} is

$$\Delta S_{\text{surroundings},1} = -\frac{Q_1}{T_{\text{sur}}} = -\frac{nRT \ln(V_f/V_i)}{T + \Delta T}.$$

The total entropy generation for case 1, combining the system and surroundings, is

$$S_{\text{gen},1} = \Delta S_{\text{system}} + \Delta S_{\text{surroundings},1} = nR \ln(V_f/V_i) - \frac{nRT \ln(V_f/V_i)}{T + \Delta T}.$$

Since $T < T + \Delta T$, the entropy generation $S_{\text{gen},1}$ is positive, reflecting the irreversibility due to the temperature difference.

Case 2: Isothermal process with heat transfer and shaft work
In the second scenario, in addition to heat transfer, a shaft rotates within the gas, performing mechanical work W_{shaft} on the gas, which is fully dissipated as frictional heat within the system. This dissipative work supplements the heat required from the surroundings.

According to the first law of thermodynamics, we have

$$\Delta U = Q_2 + W + W_{\text{shaft}} = 0.$$

Thus

$$Q_2 = -(W + W_{\text{shaft}}) = nRT \ln\left(\frac{V_f}{V_i}\right) - W_{\text{shaft}}.$$

Since W_{shaft} is positive, $Q_2 < Q_1$, meaning less heat is required from the surroundings due to the additional work from the shaft.

The entropy change for the system remains the same:

$$\Delta S_{\text{system}} = nR \ln\left(V_f/V_i\right).$$

For the surroundings, the entropy change due to the reduced heat transfer Q_2 is

$$\Delta S_{\text{surroundings},2} = -\frac{Q_2}{T_{\text{sur}}} = -\frac{nRT \ln\left(\frac{V_f}{V_i}\right) - W_{\text{shaft}}}{T + \Delta T}.$$

Comparison of entropy generation
The total entropy generation for case 2 is

$$S_{\text{gen},2} = \Delta S_{\text{system}} + \Delta S_{\text{surroundings},2} = nR \ln(V_f/V_i) - \frac{nRT \ln\left(\frac{V_f}{V_i}\right) - W_{\text{shaft}}}{T + \Delta T}.$$

Since $Q_2 < Q_1$, we find

$$-\frac{Q_2}{T_{\text{sur}}} > -\frac{Q_1}{T_{\text{sur}}}.$$

Thus

$$S_{\text{gen},2} > S_{\text{gen},1}.$$

This analysis demonstrates that the inclusion of shaft work in case 2 results in greater total entropy generation compared to the heat-only scenario in case 1, due to additional irreversibilities introduced by the dissipative work. By assuming the surroundings are at a slightly higher temperature than the system, the entropy generation is more accurately represented.

5.4.4 Adiabatic process with irreversibility (no heat transfer)

In an adiabatic process, there is no heat exchange between the system and its surroundings ($Q = 0$). Any change in the system's internal energy must come solely from the work done on or by the system. We will examine two cases involving an ideal gas: (i) an ideal, reversible adiabatic process with no entropy generation, and (ii) an adiabatic process with irreversibilities introduced by a dissipative piston.

Case 1: Ideal (reversible) adiabatic process with only expansion/compression work
In the first case, the gas undergoes an ideal, reversible adiabatic process where the work W_1 done on or by the system is the sole contributor to the change in internal energy. According to the first law of thermodynamics:

$$\Delta U = Q + W_1.$$

Since the process is adiabatic ($Q = 0$), this simplifies to

$$\Delta U = W_1.$$

For an ideal gas undergoing an adiabatic process, the relationship between pressure and volume can be described by

$$pV^\gamma = \text{constant},$$

where $\gamma = c_p/c_v$ is the heat capacity ratio. For an adiabatic expansion, the work W_1 done by the gas is negative, leading to a decrease in internal energy and a drop in temperature. For an adiabatic compression, the work W_1 done on the gas is positive, increasing the internal energy and raising the gas temperature.

Since this process is reversible, there are no irreversibilities, and the total entropy change for the system and surroundings is zero:

$$\Delta S_{\text{system}} = 0 \ \text{ and } \ \Delta S_{\text{total}} = 0.$$

With no heat exchange, the surroundings experience no entropy change, and the total entropy generation for this process is zero.

Case 2: Adiabatic process with a dissipative piston (irreversible work)
In the second scenario, the gas undergoes an adiabatic expansion or compression, but with a dissipative piston that introduces friction as it moves. This frictional force dissipates part of the work $W_{\text{dissipative}}$ done by or on the gas, causing an *increase in the internal energy* of the gas, which leads to entropy generation within the system and represents an irreversible process.

According to the first law of thermodynamics,

$$\Delta U = Q + W_2 + W_{\text{dissipative}}.$$

Since $Q = 0$ for an adiabatic process, this simplifies to

$$\Delta U = W_2 + W_{\text{dissipative}}.$$

The dissipative work $W_{\text{dissipative}}$ effectively adds to the internal energy change of the gas. For an adiabatic expansion with a dissipative piston, the temperature of the gas does not drop as much as it would in the reversible case because the dissipative work partially offsets the cooling effect of the expansion. Conversely, for an adiabatic compression with a dissipative piston, the gas temperature increases more than in the reversible case due to the additional frictional contribution to internal energy.

The key point here is that, because of the irreversibility introduced by the dissipative piston, the *end state in case 2 cannot match the end state in case 1* for the same amount of work done by or on the gas. The presence of friction changes the internal energy distribution, resulting in a different temperature and pressure compared to the reversible adiabatic process.

The entropy change of the system in case 2 is now positive, as the dissipative piston introduces irreversibilities,

$$\Delta S_{\text{system}} = S_{\text{gen,2}} > 0.$$

Since there is no heat exchange, the surroundings experience no entropy change, and the total entropy generation is entirely within the system:

$$\Delta S_{\text{total}} = S_{\text{gen,2}} > 0.$$

Comparison of entropy generation
In case 1, the total entropy change ΔS_{total} is zero, as it represents an ideal, reversible adiabatic process with no irreversibilities. In case 2, the additional dissipative work due to friction in the piston results in greater entropy generation within the system, represented by $S_{\text{gen},2}$. Thus

$$S_{\text{gen},2} > 0.$$

Furthermore, because of this dissipative work, *the final state in case 2 differs from that in case 1*. The irreversibilities prevent reaching the same pressure, volume, and temperature as in the ideal, reversible case, even if the initial states were identical.

This example illustrates how introducing dissipative work in an adiabatic process increases total entropy generation and results in a different final state than in the ideal, reversible case. Readers are encouraged to plot the adiabatic expansion and compression processes on p-V and T-S diagrams to visualize the impact of irreversibilities on the process path and final state.

Additionally, in the isobaric and isothermal examples discussed earlier, a similar analysis can be conducted by considering a lossy piston instead of shaft work, as the friction in the piston would also introduce irreversibilities. While we have primarily analysed expansion processes, readers are encouraged to apply a similar approach to compression processes to explore how irreversibilities affect entropy generation and final states in those scenarios as well.

5.5 Transition to non-quasistatic processes

In the study of thermodynamics, we have focused primarily on **quasistatic processes** —those that proceed slowly enough for the system to remain in equilibrium at every step. In a quasistatic process, the state variables (pressure, volume, temperature) are well-defined at each instant, allowing the system to follow a continuous path of equilibrium states. This assumption of equilibrium simplifies the analysis, as we can use precise thermodynamic relationships to describe the process.

However, in real-world applications, many processes are inherently **non-quasistatic**, or **non-equilibrium**, as they occur too rapidly or involve too many internal interactions for the system to maintain equilibrium throughout. In a non-quasistatic process, variables such as pressure and temperature can vary significantly within different parts of the system, making it impossible to describe the system with a single, uniform state at any given moment. These processes introduce new complexities: they require us to account for internal gradients, rapid changes, and potentially chaotic interactions, and they often result in irreversible energy dissipation and entropy generation.

In non-quasistatic processes, the departure from equilibrium means that classical thermodynamic relationships—such as those for work, heat, and entropy—cannot always be applied in their standard forms. Instead, analysing non-quasistatic processes often requires more advanced approaches, such as non-equilibrium thermodynamics.

As we transition to examining non-quasistatic processes, we will explore the unique challenges they pose and investigate the principles that govern thermodynamic behavior in far-from-equilibrium conditions. This shift offers a broader, more realistic view of how thermodynamic systems operate in nature and technology, expanding our understanding beyond the idealized, reversible framework of quasistatic processes.

5.6 Introduction to non-quasistatic processes and moving boundary work transfer for an ideal gas

In previous chapters, we discussed **quasistatic processes**—those in which the system remains in mechanical equilibrium with its surroundings at each step. This equilibrium allows us to define the system's pressure throughout the process. However, in a **non-quasistatic process**, the system is not in mechanical equilibrium with its surroundings and, as a result, the pressure within the system varies in ways that are not well-defined at each instant. Consequently, non-quasistatic processes cannot be represented on standard thermodynamic property diagrams, such as the pressure–volume (p–V) or temperature–entropy (T–S) diagrams. However, we can still determine changes in state variables, such as internal energy and entropy, by comparing the initial and final states, as these are state functions and do not depend on the process path.

5.6.1 Moving boundary work transfer in a non-quasistatic process

In a non-quasistatic process, **moving boundary work** occurs when an external force acts to change the system's volume (i.e. it moves the system boundary). The work done on the system due to this external force is

$$W = \vec{F}_{\text{ext}} \cdot d\vec{s} = (p_{\text{ext}} \cdot A)d\left(\frac{V}{A}\right) = -p_{\text{ext}}\,dV,$$

where:
- p_{ext} is the **external pressure**, defined as F_{ext}/A,
- dV is the **volume change** equal to $\vec{s} \cdot A$, and
- the minus sign accounts for the convention that work done on the system is positive when the volume decreases.

In a quasistatic process, the system is always in equilibrium with its surroundings, meaning that the system pressure p equals the external pressure p_{ext} at all points during the process. Thus, the work expression simplifies to

$$\text{Quasistatic process:} \quad W = -p_{\text{ext}}\,dV = -p\,dV.$$

However, in a non-quasistatic process, the system is not in mechanical equilibrium with its surroundings, so the external pressure p_{ext} differs from the system pressure p. Therefore, for non-quasistatic processes, the work transferred to the system is

$$\text{Non-quasistatic process:} \quad W = -p_{\text{ext}}\,dV.$$

In this case, we must use the external pressure p_{ext} rather than the system pressure p to calculate the boundary work, as the system's pressure is not well-defined during the non-equilibrium process.

5.6.1.1 Examples of moving boundary work in non-quasistatic processes

1. **Gas expansion in a vacuum (external pressure zero)**

   Consider an ideal gas enclosed in a cylinder with a weightless piston, initially held in place by stops (figure 5.1). The initial pressure and volume of the gas are p_1 and V_1, respectively. When the stops are removed, the gas expands until it reaches a second set of stops, increasing to volume V_2 with a final pressure p_2. Since the external pressure $p_{\text{ext}} = 0$ (a vacuum),

$$W = -\int p_{\text{ext}}\, dV = 0.$$

   Although the gas volume increases, no work is done by the system to expand, as there is no opposing external force.

2. **Gas expansion against a constant external pressure**

   In this scenario, a mass m is placed on a weightless piston, creating a constant external pressure $p_{\text{ext}} = mg/A$, where A is the piston area (figure 5.2). When the gas expands, the moving boundary work is

$$W = -\int_{V_1}^{V_2} p_{\text{ext}}\, dV = -\int_{V_1}^{V_2} \frac{mg}{A}\, dV = -\frac{mg}{A}(V_2 - V_1).$$

   If the mass m is chosen such that $mg = p_2 A$, the piston reaches the second set of stops, with the work done by the gas given by

$$W = -p_2(V_2 - V_1).$$

   This area on a p-V diagram represents the work performed by the gas (figure 5.2 (bottom)).

3. **Multi-step expansion with incremental mass changes**

   Now, suppose we add another mass m on the piston (figure 5.3), creating a higher external pressure p_3. After removing the stops, the gas expands to volume V_3. Then, we remove one mass, allowing the gas to further expand to

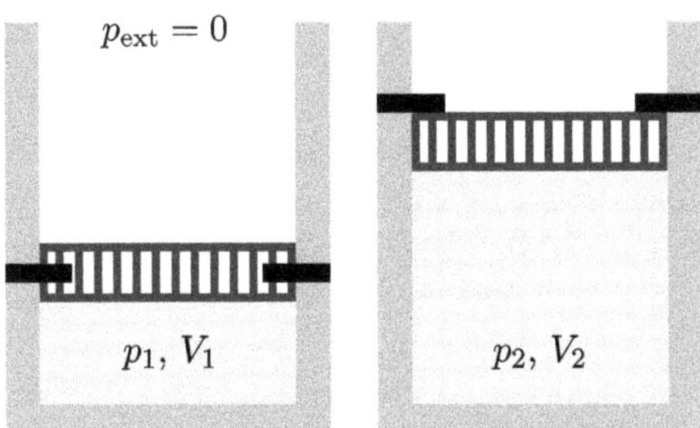

Figure 5.1. An ideal gas enclosed in a cylinder with a weightless piston. Left: The initial state with the piston held by stops. Right: The final state with the stops removed and the gas expanded.

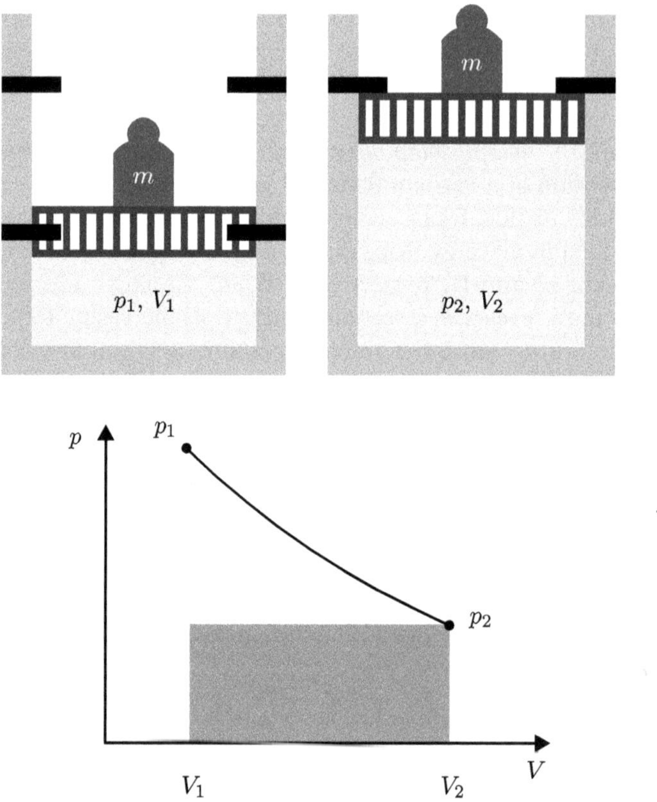

Figure 5.2. Top left: The initial state, a mass m placed on a weightless piston held by stops. Top right: The final state after the bottom stops are removed and the gas has expanded. Bottom: The pV diagram.

volume V_2. The work done during this process can be broken down as follows:

$$W = - \int_{V_1}^{V_3} p_{\text{ext}}\, dV - \int_{V_3}^{V_2} p_{\text{ext}}\, dV = -\int_{V_1}^{V_3} p_3\, dV - \int_{V_3}^{V_2} p_2\, dV$$
$$= - p_3(V_3 - V_1) - p_2(V_2 - V_3).$$

The shaded areas in the bottom panel of figure 5.3 represent the work performed by the gas, which is greater than that with a single mass.

4. **Continuous expansion with incrementally smaller masses**

If we place incrementally smaller masses on the piston and remove them one by one (figure 5.4), the total work done by the gas approaches that of a quasistatic process, represented by the bold line in the p-V diagram.

To conclude: In the limit of small changes to the external force, the work performed by the gas converges to that of a quasistatic process.

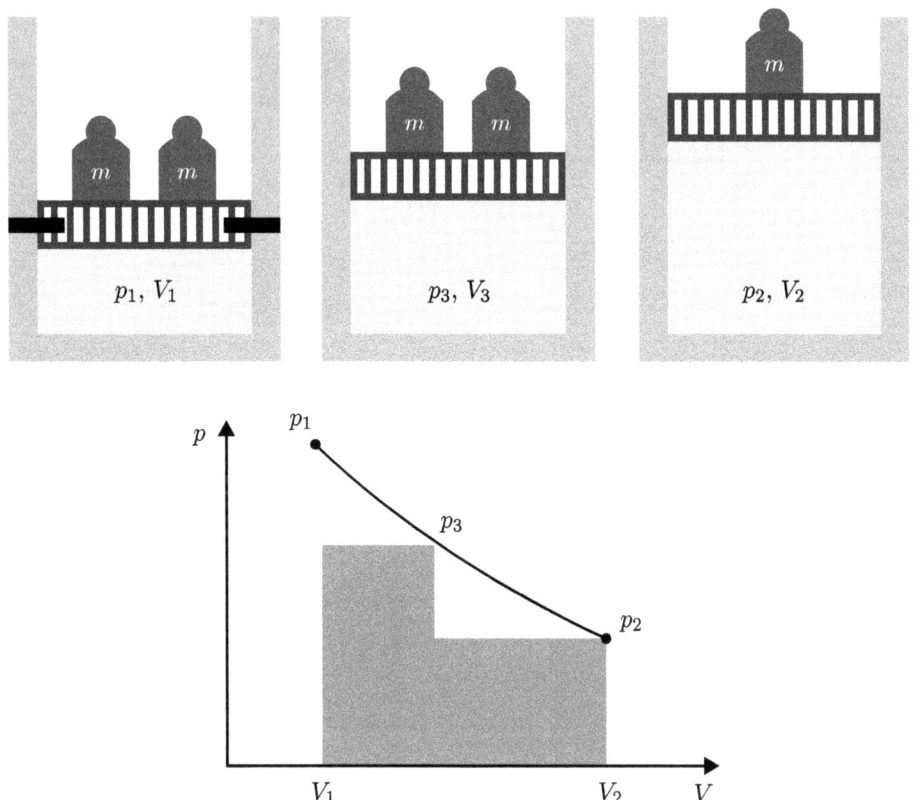

Figure 5.3. Top left: The inital state with two masses m on the piston which is held in place by stops. Top middle: The second state after the stops are removed. Top right: The final state with one mass removed. Bottom: The pV diagram.

5.6.1.2 Suggested analysis for compression processes

The above examples illustrate moving boundary work during gas expansion under various external conditions. Readers are encouraged to conduct a similar analysis for compression processes, starting from volume V_2 to V_1. This analysis could involve a single mass, two masses, or a series of smaller masses to observe the effects of different external pressures during compression. By applying the same systematic approach, readers can explore how external forces influence work in non-quasistatic compression processes.

5.6.2 Non-quasistatic processes

This section introduces a methodology for calculating changes in internal energy and entropy in non-quasistatic processes, focusing on ideal gas behavior without additional forms of work transfer. The first and second laws of thermodynamics still apply to non-quasistatic processes. For clarity, we restate them here:

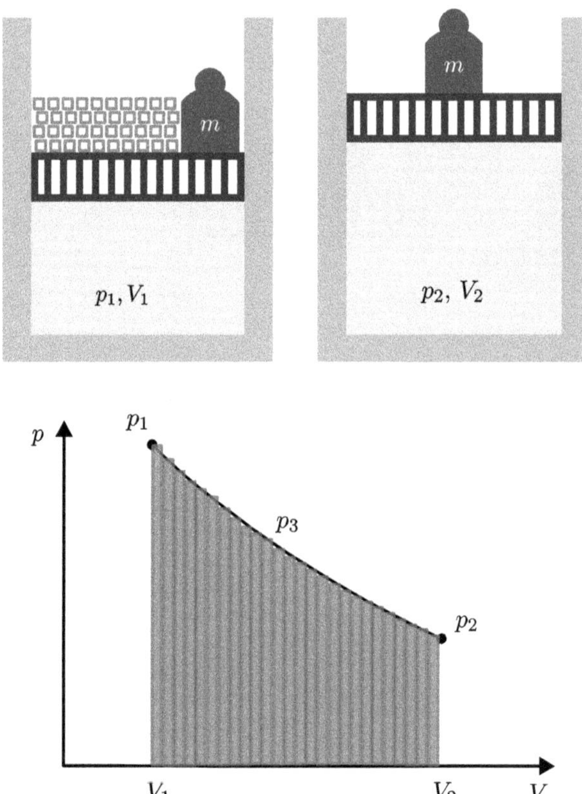

Figure 5.4. Top: Incrementally smaller masses on the piston. Bottom: The pV diagram.

- **First law** (with only pV work):

$$\Delta U = Q + W.$$

- **Second law**:

$$\Delta S = \int \frac{Q}{T} + S_{\text{irr}},$$

where S_{irr} represents the irreversible entropy generated within the system due to irreversibilities.

We will analyse two types of non-quasistatic processes for an ideal gas: an adiabatic expansion and an isothermal compression, and we will compare each with its quasistatic counterpart.

5.6.2.1 Adiabatic expansion—quasistatic versus non-quasistatic
Consider an ideal gas initially at pressure p_1 and volume V_1, enclosed in a cylinder with a weightless piston (figure 5.5).

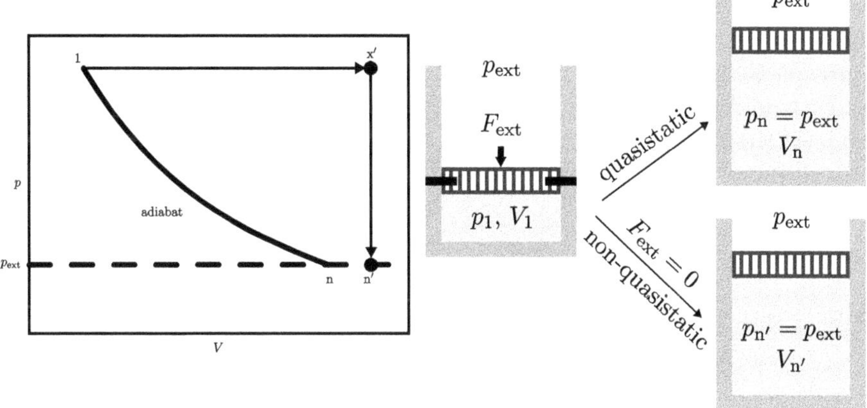

Figure 5.5. Schematic representation of a quasistatic versus non-quasistatic adiabatic expansion of an ideal gas in a piston–cylinder arrangement. In the quasistatic process, the external force is reduced gradually, allowing the system to follow an equilibrium adiabat. In contrast, the non-quasistatic expansion occurs rapidly against a constant external pressure, leading to irreversibilities and greater final volume due to entropy generation.

Quasistatic adiabatic expansion

In a quasistatic adiabatic expansion, an external force F_{ext} on the piston is gradually reduced, allowing the gas to expand along an adiabat. At the final state n, the gas pressure equals the external pressure. For a quasistatic adiabatic process:

$$pV^{\gamma} = p_1 V_1^{\gamma} = p_n V_n^{\gamma} = \text{constant}$$

Where γ is the heat capacity ratio $\frac{c_p}{c_v}$. The ideal gas law applies throughout:

$$\frac{pV}{T} = \frac{p_1 V_1}{T_1} = \frac{p_n V_n}{T_n} = \text{constant}$$

Since this is an adiabatic process, $Q = 0$. Applying the first law:

$$\Delta U = W$$

With no irreversible entropy generation ($S_{irr} = 0$), the entropy change is:

$$\Delta S = \int \frac{Q}{T} + S_{irr} = 0$$

Consider an ideal diatomic gas initially at pressure $p_1 = 10$ bar, volume $V_1 = 1$ L, and temperature $T_1 = 300$ K. The gas is expanded to a pressure of 1 bar. For this quasistatic expansion from state 1 to state n, we calculate:
- Change in internal energy: $\Delta U = -1205$ J.
- Work done: $W = -1205$ J.
- Entropy change: $\Delta S = 0$.

Non-quasistatic adiabatic expansion

Now consider a non-quasistatic expansion starting from the same initial state. In this process, the gas expands rapidly against the external pressure p_{ext}, so the system is not in equilibrium with its surroundings, and intermediate states are not well-defined. At the final state n', the gas pressure equals the external pressure, but the volume differs from the quasistatic case.

The moving boundary work in the non-quasistatic process is:

$$W = - p_{ext}\left(V_{n'} - V_1\right)$$

With $Q = 0$, applying the first law yields:

$$\Delta U = W = U_{n'} - U_1 = - p_{ext}\left(V_{n'} - V_1\right)$$

To evaluate the final volume $V_{n'}$, we use the thermodynamic expression for internal energy of an ideal gas:

$$U = \frac{f}{2}nRT = \frac{f}{2}pV$$

This relation allows us to express the change in internal energy as:

$$\Delta U = \frac{f}{2}\left(p_{n'} V_{n'} - p_1 V_1\right)$$

Equating this with the expression from the first law:

$$\frac{f}{2}\left(p_{n'} V_{n'} - p_1 V_1\right) = - p_{ext}\left(V_{n'} - V_1\right)$$

Rearranging terms gives:

$$\left(\frac{f}{2}p_{n'} + p_{ext}\right)V_{n'} = \frac{f}{2}p_1 V_1 + p_{ext} V_1$$

Substituting the known values:
- $p_1 = 10 \text{ bar} = 10^6 \text{Pa}$
- $V_1 = 1L = 10^{-3}\text{m}^3$
- $p'_n = p_{ext} = 1 \text{ bar} = 10^5$
- $f = 5$

$$\left(\frac{5}{2}\cdot 10^5 + 10^5\right)V_{n'} = \frac{5}{2}\cdot 10^6 \cdot 10^{-3} + 10^5 \cdot 10^{-3}$$

$$\left(2.5 \cdot 10^5 + 10^5\right)V_{n'} = \left(2.5 + 0.1\right)\cdot 10^3$$

$$\left(3.5 \cdot 10^5\right)V_{n'} = \left(2.6\right)\cdot 10^3$$

$$V_{n'} = \frac{2.6 \cdot 10^3}{3.5 \cdot 10^5} = 7.43 \cdot 10^{-3} \text{m}^3 = 7.43 \text{L}$$

Now that we have the final volume, we can compute the internal energy:

$$U_{n'} = \frac{f}{2} p_{n'} V_{n'} = \frac{5}{2} \cdot 10^5 \cdot 7.43 \cdot 10^{-3} = 1857 \text{J}$$

Since the initial internal energy was $U_1 = 2500$ J, we find:

$$\Delta U = U_{n'} - U_1 = -643 \text{J}$$

and hence:

$$W = -643 \text{J}$$

To compute the entropy change, we consider a hypothetical reversible path from state 1 to n' consisting of:

1. An **isobaric** process from 1 to x': $p = p_1 \, V_1 \rightarrow V_{n'}$
2. An **isochoric** process from x' to n': $V = V_{n'}$, $p_1 \rightarrow p_{n'}$

Using:

$$\Delta S = n c_p \ln\left(\frac{T_{x'}}{T_1}\right) + n c_v \ln\left(\frac{T_{n'}}{T_{x'}}\right)$$

From the ideal gas law:

$$T_{x'} = \frac{p_1 V_{n'}}{nR}, \; T_{n'} = \frac{p_{n'} V_{n'}}{nR}$$

Substituting these gives the total entropy change, which evaluates numerically to:

$$\Delta S = 4.2 \frac{\text{J}}{\text{K}}$$

This value represents the entropy generated due to irreversibility, since the actual adiabatic path is not reversible. The increase in entropy and volume $V_{n'} > V_n$ highlights the thermodynamic consequences of performing the expansion non-quasistatically.

Table 5.1. Thermodynamic state variables for an ideal diatomic gas undergoing adiabatic expansion—comparison between quasistatic and non-quasistatic processes. The table lists pressure, volume, internal energy, and temperature for initial state, final states (quasistatic and non-quasistatic), and intermediate reference states used for entropy calculations.

State	State 1	State n	State n'	State x'
p (bar)	10	1	1	10
V (L)	1	5.18	7.43	7.43
U (J)	2500	1295	1857	18571
T (K)	300	155	223	2229

Main findings for adiabatic expansion comparison
- **Work comparison**: The magnitude of W in the non-quasistatic expansion process is lower than in the quasistatic process.
- **Volume difference**: In the non-quasistatic process, the final volume $V_{n'}$ is greater than the quasistatic final volume V_n, as irreversible entropy generation requires a larger volume change.
- **Entropy generation**: The entropy change in the non-quasistatic process is positive due to irreversible entropy generation. In contrast, the entropy change for the quasistatic process is zero, making it reversible.

This example clarifies how entropy generation and work differ in quasistatic versus non-quasistatic adiabatic expansions. By following alternative paths (involving states x and x') for entropy calculations, we can compare these processes with a more systematic approach. To deepen your understanding, you are encouraged to conduct a similar analysis for an **adiabatic compression process**, considering both quasistatic and non-quasistatic scenarios. Start with the same initial conditions and examine how changes in internal energy, work done, and entropy differ between the two cases. By following a similar methodology—perhaps using an alternative path with intermediate states as we did with states x and x'—you will gain insights into the effects of irreversibility on entropy generation and final state differences during compression. This exercise will further illustrate the unique characteristics of non-quasistatic processes.

5.6.2.2 Isothermal compression—quasistatic versus non-quasistatic
Now let us examine an ideal gas compression process under isothermal conditions, contrasting a quasistatic process with a non-quasistatic one.

5.6.2.2.1 Quasistatic isothermal compression
The gas is initially at a pressure of 1 bar, a volume of 1.0 l, and a temperature of 300 K. It is compressed isothermally to a final volume of 0.1 l, with the surroundings as a constant-temperature reservoir at 290 K.

The work done in the quasistatic isothermal process is

$$W = -\int_1^n p\,dV = -nRT_0 \ln \frac{V_f}{V_i} = -p_i V_i \ln \frac{V_f}{V_i} = 100 \ln 10 \text{ J}.$$

With no change in internal energy ($\Delta U = 0$) for an isothermal process, the heat transfer is

$$Q = -W = -100 \ln 10 \text{ J}.$$

The entropy change of the system is

$$\Delta S_{\text{system}} = \int \frac{Q}{T} = \frac{-100 \ln 10}{300} = -\frac{\ln 10}{3} \text{ J K}^{-1}.$$

Since the surroundings act as a heat reservoir, the entropy change is

$$\Delta S_{\text{surroundings}} = \frac{-Q}{T_{\text{reservoir}}} = \frac{100 \ln 10}{290} = \frac{300 \ln 10}{3 \times 290} \text{ J K}^{-1}.$$

Thus, the total entropy change is

$$\Delta S_{\text{total}} = \Delta S_{\text{system}} + \Delta S_{\text{surroundings}} = \frac{\ln 10}{3}\left(\frac{300}{290} - 1\right) > 0.$$

5.6.2.2.2 *Non-quasistatic isothermal compression*
In this case, an external force exerts a constant pressure of 10 bar on the gas, causing an instantaneous compression to 0.1 l. The final temperature is maintained at 300 K.
   The work done in the non-quasistatic process is

$$W = -p_{\text{ext}}(V_f - V_i) = -10 \text{ bar} \times (0.9 \text{ l}) = 900 \text{ J}.$$

With no change in internal energy ($\Delta U = 0$), the heat transfer is

$$Q = -W = -900 \text{ J}.$$

The entropy change of the system is the same as in the quasistatic process:

$$\Delta S_{\text{system}} = -\frac{\ln 10}{3} \text{ J K}^{-1}.$$

However, the entropy change of the surroundings differs due to the larger heat transfer:

$$\Delta S_{\text{surroundings}} = \frac{-Q}{T_{\text{reservoir}}} = \frac{900}{290} \text{ J K}^{-1}.$$

The total entropy change for the non-quasistatic process is

$$\Delta S_{\text{total}} = \Delta S_{\text{system}} + \Delta S_{\text{surroundings}} = -\frac{\ln 10}{3} + \frac{900}{290} > 0.$$

This larger total entropy change in the non-quasistatic process indicates higher irreversibility than in the quasistatic case.

5.6.2.2.3 *Main findings*
- **Work comparison**: The magnitude of W in non-quasistatic compression processes is generally higher than in quasistatic processes.
- **Entropy generation**: Entropy change in the non-quasistatic process is positive due to irreversible entropy generation, whereas the total entropy change for the quasistatic process is zero, making it a reversible process.

In summary, both examples demonstrate that non-quasistatic processes lead to higher entropy generation, indicating greater irreversibility compared to their quasistatic counterparts. Readers are encouraged to apply this approach to other scenarios, such as non-quasistatic adiabatic compression, as well as to explore arbitrary variations in external conditions.

It is important to note that there are countless ways in which a non-quasistatic process can unfold, as the system is not constrained to maintain equilibrium with its surroundings. Because of this, analysing a non-quasistatic process requires specific information about how the process occurred or detailed data about the states involved. Only with this information—whether through measurement, experimental data, or clear process definitions—can an accurate analysis of the system's behavior and energy transformations be performed.

5.7 Thermodynamic identity

In this final section, we introduce the **thermodynamic identity**, a fundamental relation that links internal energy, entropy, temperature, pressure, and volume in a concise mathematical form. This identity encapsulates the changes in internal energy in terms of entropy and volume, providing a powerful tool for analysing thermodynamic processes.

5.7.1 Derivation of the thermodynamic identity

We begin by considering entropy S as a function of internal energy U, volume V, and particle number N. For simplicity, we will focus on a fixed number of particles, so entropy becomes a function of U and V alone:

$$S = S(U, V).$$

Taking the differential of entropy S, we have

$$dS = \left(\frac{\partial S}{\partial U}\right)_V dU + \left(\frac{\partial S}{\partial V}\right)_U dV.$$

From our previous discussions on thermal and mechanical equilibrium, we defined **temperature** T and **pressure** p in terms of these partial derivatives:

$$\frac{1}{T} = \left(\frac{\partial S}{\partial U}\right)_V$$

$$p = T\left(\frac{\partial S}{\partial V}\right)_U.$$

Substituting these definitions into the differential of entropy, we obtain

$$dS = \frac{1}{T}dU + \frac{p}{T}dV.$$

Rearranging, we can multiply through by T to isolate dU:

$$TdS = dU + pdV.$$

Rearranging further, we arrive at the **thermodynamic identity**:

$$dU = TdS - pdV.$$

This identity provides an essential link between state variables, relating changes in **internal energy** U to changes in **entropy** S and **volume** V. It reveals that, for a small change in a system's internal energy, the amount of energy is partitioned between the change in entropy (multiplied by temperature) and the change in volume (multiplied by pressure).

5.7.2 Significance of the thermodynamic identity

The thermodynamic identity captures the fundamental interplay of energy, entropy, and volume in thermodynamic systems. Each term in the identity—TdS and $-pdV$—represents a component of the system's internal energy change:

- **Entropy-related energy change**: The term TdS represents the portion of internal energy change associated with entropy change. This includes contributions from both heat transfer and irreversible processes, such as frictional effects or dissipative work, which increase entropy within the system.
- **Work**: The term $-pdV$ corresponds to the work associated with changes in the system's volume. This is often called boundary work, as it reflects the energy required for the system to expand or compress against an external pressure.

Together, these terms describe how internal energy changes in response to both entropy and volume variations, making the thermodynamic identity a versatile tool for analysing any thermodynamic process, whether reversible or irreversible.

This identity is foundational because it holds for any infinitesimal process, whether it is reversible or irreversible, as long as it involves small changes. It allows us to analyse complex processes by breaking them down into contributions from heat and work, thereby deepening our understanding of how energy is conserved and transformed in thermodynamic systems.

5.7.3 Thermodynamic identity clarification

In deriving the thermodynamic identity, many authors take a simplified approach, starting directly from the first law of thermodynamics $dU = Q + W$ and substituting $Q = TdS$ and $W = -pdV$. While this approach is often convenient, it overlooks an important aspect of real thermodynamic systems—namely, that other forms of work transfer, such as friction and shaft work, can occur, especially in irreversible processes.

Our derivation, however, takes a more comprehensive route by incorporating the idea that *entropy change results from both heat transfer and irreversibilities*. Specifically, we start with the expression $dS = \frac{Q}{T} + S_{\text{irr}}$, where S_{irr} represents the entropy generated due to irreversible effects within the system. Thus, the internal energy change can be more completely expressed as

$$dU = Q + W_{pV} + W_{\text{other}},$$

where:

- Q represents heat transfer, leading to a change in system entropy,
- $W_{pV} = -pdV$ is the work associated with volume change (boundary work), and
- W_{other} represents additional forms of work, such as frictional or shaft work, which can contribute to entropy generation.

This perspective respects the contributions of all energy transfers and entropy changes in real-world systems, offering a more complete picture. The entropy change in this approach accounts for both the reversible entropy change from heat transfer and the irreversible contribution from other forms of work.

5.7.4 The thermodynamic identity in any process

Through this derivation, we obtain the thermodynamic identity

$$dU = TdS - pdV.$$

This identity holds universally for any infinitesimal process, regardless of whether it is reversible or irreversible, because it shows the interplay of **state variables**—internal energy U, entropy S, temperature T, pressure p, and volume V—that define the system. By deriving it through the fundamental principles of entropy and energy transfer, we ensure that it remains applicable even when additional irreversibilities and forms of work transfer are present.

This approach not only honors the first law but also provides a fuller understanding of energy and entropy changes, making it a powerful tool for analysing any thermodynamic process, no matter the complexity.

It is often a limiting perspective that, when first introduced to thermodynamics, readers may come away with the impression that it applies only to mechanical or pneumatic systems—those involving heat transfer, expansion work, and similar phenomena. But this is incorrect. The principles of thermodynamics are universal and extend well beyond mechanical systems. In many physical domains—such as electrical, magnetic, and chemical systems—additional forms of work must be considered. For instance, electrical systems may involve terms like ϕdQ (electric potential times charge), magnetic systems include HdM (magnetic field times change in magnetization), and chemical systems require terms like μdN (chemical potential times particle number). These additional contributions modify the thermodynamic identity and show that entropy can change through more than just heat transfer. Nevertheless, even in such cases, the mechanical work term pdV does not vanish. As long as the system undergoes volume changes under ambient conditions, work is done by or against the surrounding atmosphere, and this must be accounted for in the total energy balance.

5.8 Summary

This chapter provides a detailed examination of thermodynamic processes that extend beyond idealized, reversible assumptions. Through an analysis of quasistatic processes with irreversible work transfer, we highlight the essential role of dissipative forces—such as friction and inelastic deformation—in generating entropy and causing energy losses that cannot be fully recovered. By distinguishing between reversible and irreversible work components, we gain a clearer understanding of how energy distribution is affected by irreversibilities in real-world systems.

In discussing non-quasistatic processes, the chapter underscores the limitations of equilibrium-based thermodynamic analysis and introduces methodologies for evaluating systems far from equilibrium. Importantly, we have shown a clear strategy for determining changes in internal energy and entropy in non-quasistatic processes. Since these quantities are state functions, their values depend only on the initial and final states—not the path taken. This allows one to calculate entropy or internal energy changes by conceptually constructing alternate quasistatic paths, even if the actual process is highly irreversible or lacks intermediate equilibrium states.

The chapter concludes with the derivation of the thermodynamic identity, incorporating both heat transfer and irreversibilities. This identity provides a unified framework for analysing energy changes in thermodynamic systems, applicable to any infinitesimal process, whether reversible or irreversible. Through this comprehensive approach, we establish the thermodynamic identity as a powerful tool for understanding the interplay of state variables, reinforcing its foundational role in thermodynamics.

IOP Publishing

A Classical Thermodynamics Toolkit

Srinivas Vanapalli

Chapter 6

Thermodynamic cycles

6.1 Introduction

Thermodynamic cycles are at the core of many energy transformation systems, from engines that produce mechanical work to coolers that maintain low temperatures. These cycles represent closed-loop sequences of thermodynamic processes through which a working substance interacts with its surroundings. A key concept in understanding thermodynamic cycles is the distinction between a process, where a system transitions between states, and a cycle, where the system returns to its initial state.

We will begin this chapter by distinguishing between a thermodynamic process and a thermodynamic cycle. This distinction will be illustrated using simple real-world examples. We will then explore why thermodynamic cycles are so important in practice, especially in energy conversion systems. From there, we will introduce the classification of thermodynamic cycles into power cycles, refrigeration cycles, and heat pump cycles. We will focus primarily on closed systems and pure substances, although the same principles apply more broadly. Finally, we will examine a few representative idealized cycles to illustrate the underlying principles of thermodynamic analysis.

This chapter bridges the theoretical framework of thermodynamic processes discussed earlier with real-world systems.

6.2 The difference between a process and a cycle

In thermodynamics, it is essential to understand the difference between a *process* and a *cycle* to grasp how energy transformation and efficiency work within systems. This distinction is fundamental because processes and cycles serve unique roles in managing energy flow, achieving specific goals such as cooling, heating, or work generation.

doi:10.1088/978-0-7503-6029-6ch6   6-1

6.2.1 The definition of a thermodynamic process

A **thermodynamic process** refers to a change in the state of a system from an initial state to a final state. This can involve changes in variables such as pressure, volume, and temperature, which occur as the system undergoes energy interactions (heat transfer or work). Importantly, a process is not necessarily closed or repeatable—it can happen only once, and the system might not return to its original state.

6.2.2 Definition of a thermodynamic cycle

A **thermodynamic cycle**, on the other hand, is a sequence of processes that ultimately returns a system to its original state. In other words, in a cycle, the initial and final states are the same. This closed-loop sequence allows for continuous operation, making cycles essential in applications such as engines and refrigeration systems where consistent performance over time is required. Cycles enable a system to repeatedly transform energy, often converting heat into work or vice versa.

6.2.3 Examples to differentiate a process and a cycle

- **Air conditioning in an airplane (process)**: Air conditioning in an airplane is an example of an **open system** that undergoes a thermodynamic process but does not complete a closed cycle. The air conditioning system relies on compressed air to cool the cabin, especially when the aircraft is on the ground. Here is how the process typically works:
    1. *Compressed air supply*: When the aircraft is on the ground, an auxiliary power unit (APU) provides compressed air, which is directed into an air conditioning unit to initiate the cooling process.
    2. *Primary heat exchange*: The high-pressure, high-temperature air flows through a primary heat exchanger, where it loses heat to ambient air, cooling down but remaining under relatively high pressure.
    3. *Expansion for cooling*: This cooled, compressed air then passes through a turbine, where it expands, causing a sharp drop in temperature. This expansion produces the cold air needed for cabin cooling.
    4. *Continuous one-way flow*: This cold air is then distributed throughout the cabin and eventually vented out. Because this air is not recycled back to its original state, the air conditioning system is considered an open system performing a thermodynamic *process* rather than a cycle.
- **Brayton cycle (closed cycle)**: In contrast, the **Brayton cycle** is a thermodynamic cycle commonly used in closed-loop applications, such as gas turbines and some advanced refrigeration systems. The closed Brayton cycle consists of a series of processes—compression, heat exchange, and expansion—that allow the system to return to its initial state by the end of each cycle. Here is how it operates:
    1. *Compression*: The working fluid (often air or a gas) is compressed, which raises its pressure and temperature.

2. *Heat exchange*: The compressed air then flows through a heat exchanger, where it releases some of its heat to the surroundings. This stage reduces the temperature of the air while it remains under high pressure.
3. *Expansion*: The cooled, compressed air is then expanded through a turbine, where it drops in both temperature and pressure, generating the desired cooling effect.
4. *Closed-loop recirculation*: The low-pressure, cold air is recirculated back to the compressor, where it repeats the cycle. In this closed-loop configuration, the Brayton cycle can continuously perform cooling or power generation, depending on the set-up.

This closed Brayton cycle is distinct from the one-way flow in an airplane air conditioning system because the working fluid undergoes a continuous loop, returning to its original state after each cycle. This closed-loop operation allows for sustained energy transformation, highlighting the difference between an ongoing **process** and a repeatable thermodynamic **cycle**.

6.3 Why do we need cycles?

In thermodynamics, cycles serve as a foundational concept for understanding how energy is transformed and utilized efficiently. The primary purpose of cycles is to facilitate the conversion of energy from one form to another. This is particularly important in applications where heat, a disorganized form of energy, needs to be transformed into useful work or transferred efficiently for heating or cooling purposes.

A thermodynamic cycle consists of a sequence of processes that return a system to its original state. By completing this cycle, energy can be extracted or supplied repeatedly without permanently altering the working substance. This is critical in engineering systems that need continuous operation, such as engines, refrigerators, and power plants.

The essence of a cycle lies in its ability to exploit the inherent properties of materials, such as gases, liquids, or mixtures, to achieve this transformation. For instance:

- In **power generation**, cycles such as the Brayton or Rankine cycles convert thermal energy from fuel combustion or solar heating into mechanical work, which is then turned into electricity.
- In **refrigeration and heat pumps**, cycles such as the vapor-compression cycle utilize work input to transfer heat from a cooler region to a hotter one, enabling cooling or heating, respectively.
- Advanced technologies, such as **magnetic refrigerators** or **dilution fridges**, employ specialized cycles to achieve extremely low temperatures for scientific or industrial purposes.

Cycles are not only essential for energy conversion but also for optimizing efficiency. They allow engineers to design systems that maximize the useful output from a given energy input while minimizing losses. The Carnot cycle, for example, provides an idealized benchmark for the maximum efficiency a heat engine can achieve between two temperatures.

6.4 Examples of thermodynamic cycles across applications

Thermodynamic cycles are the backbone of numerous technologies, enabling energy conversion and heat transfer for a wide range of applications. Below are examples that illustrate how different cycles are utilized across various domains.

6.4.1 Solar power plant: using cycles for renewable energy generation

In solar power plants, thermodynamic cycles such as the **Rankine cycle** or **Brayton cycle** play a crucial role in converting solar energy into electricity. Concentrated solar power (CSP) systems, for example, use mirrors or lenses to focus sunlight onto a heat transfer fluid. The thermal energy is then used to produce steam, driving a turbine connected to an electric generator. These cycles ensure efficient energy conversion while promoting sustainability by utilizing an abundant and clean energy source.

6.4.2 Home refrigerator: an example of a refrigeration cycle for cooling

The **vapor-compression refrigeration cycle** is fundamental to modern refrigeration systems found in homes. This cycle transfers heat from the interior of the refrigerator to the surrounding environment, keeping food and beverages cool. Key processes involve compression, condensation, expansion, and evaporation, where a refrigerant circulates through the system, absorbing heat inside the refrigerator and releasing it outside. The cycle operates efficiently to maintain the desired temperature with minimal energy consumption.

6.4.3 Heat pump: cycle application for heating purposes

A **heat pump** is essentially a refrigeration cycle in reverse, designed to move heat from a cooler region to a warmer one. In cold climates, heat pumps efficiently extract heat from outdoor air (or other sources such as the ground) and deliver it indoors for heating purposes. By using a thermodynamic cycle similar to the vapor-compression cycle, heat pumps provide a sustainable and energy-efficient alternative to traditional heating methods, reducing greenhouse gas emissions.

6.4.4 Magnetic refrigerator: non-gas-based cooling technology

Magnetic refrigeration relies on the **magnetocaloric effect** instead of conventional gases. In this cycle, a magnetic material is cyclically magnetized and demagnetized, causing changes in its temperature. The material absorbs heat from the space to be cooled during demagnetization and releases heat during magnetization. Magnetic refrigerators offer a promising alternative to traditional refrigeration systems, with the potential for higher efficiency and the elimination of harmful refrigerants.

6.4.5 Dilution refrigerator: used in ultra-low-temperature physics

The **dilution refrigerator** is a specialized device used in scientific research to achieve temperatures close to absolute zero. This technology operates on the principle of phase separation in a mixture of helium-3 and helium-4 isotopes. The cycle involves continuous dilution of helium-3 into helium-4 at ultra-low temperatures, absorbing heat and maintaining the desired low temperature. Dilution refrigerators are indispensable for experiments in quantum computing, superconductivity, and other fields requiring precise control over extreme cryogenic conditions.

6.5 Classification of thermodynamic cycles

Thermodynamic cycles can be broadly categorized based on their primary function. While the underlying principles remain similar, the objectives of these cycles determine their design and operational parameters. Two major categories are power cycles and refrigeration cycles.

6.5.1 Power cycles: designed to produce work

Power cycles are engineered to convert heat into mechanical work, which is often further transformed into electrical energy. These cycles are integral to the functioning of power plants and other energy generation systems.

Key examples include:

- **Rankine cycle**: This is the backbone of most thermal power plants, where steam is generated by heating water, used to drive a turbine, and then condensed back into water to complete the cycle. It is commonly employed in coal, nuclear, and solar thermal power plants.
- **Brayton cycle**: Found in gas turbines and jet engines, this cycle involves compressing air, heating it by fuel combustion, and then expanding the hot gases through a turbine to produce work.
- **Otto and Diesel cycles**: These are internal combustion engine cycles used in automobiles. The Otto cycle is characteristic of gasoline engines, while the Diesel cycle is used in diesel engines.

Power cycles are designed to maximize efficiency, often operating between high- and low-temperature reservoirs. The efficiency is fundamentally limited by the laws of thermodynamics, with the Carnot cycle serving as an idealized benchmark.

6.5.2 Refrigeration cycles: designed to transfer heat

Refrigeration cycles, in contrast, focus on transferring heat from a lower-temperature region to a higher-temperature one, often against the natural direction of heat flow. These cycles rely on work input to achieve their objectives, making them essential for cooling and heating applications.

Key examples include:

- **Vapor-compression refrigeration cycle**: This is the most widely used refrigeration cycle, found in home refrigerators, air conditioners, and heat pumps. It employs a refrigerant that circulates through compression, condensation, expansion, and evaporation processes to move heat.
- **Absorption refrigeration cycle**: This cycle uses a heat source instead of mechanical work for driving the refrigeration process. It is often employed in industrial and off-grid applications.
- **Reverse Brayton cycle**: A variation of the Brayton cycle used for cooling, especially in cryogenic systems and aerospace applications.

While power cycles are aimed at extracting useful work from heat, refrigeration cycles are designed to control thermal conditions by absorbing or rejecting heat. Together, these two classifications encompass the majority of practical thermodynamic cycles, enabling diverse energy and thermal management applications.

6.6 Cycle performance

In a thermodynamic cycle, the working substance—the system—undergoes a series of processes that eventually return it to its initial state. Previously, we examined processes and energy interactions between the system and its surroundings. In cycles, multiple processes can occur simultaneously across different parts of the working substance. For instance, while one part of the substance absorbs heat, another part may be rejecting heat or performing work. This simultaneous occurrence is characteristic of most thermodynamic cycles, where the system as a whole continuously evolves through the cycle. However, in time-separated cycles, such as magnetocaloric cycles, the entire system undergoes one process at a time in sequence. This distinction helps us understand how the working substance interacts with surroundings throughout a cycle, which we will explore in detail in this section.

We will use this simple four process in a cycle shown on a property diagram (figure 6.1) to discuss the thermodynamics of a cycle.

In a thermodynamic cycle, the working fluid undergoes a series of processes and ultimately returns to its initial state. This cyclic behavior ensures that the net change in state properties, such as pressure, volume, internal energy, and entropy, over a complete cycle is zero:

$$\Delta h_{12} + \Delta h_{23} + \Delta h_{34} + \Delta h_{41} = 0, \quad \Delta s_{12} + \Delta s_{23} + \Delta s_{34} + \Delta s_{41} = 0.$$

From the first law of thermodynamic for a closed system, the total energy change over the cycle is

$$\Delta h_{12} + \Delta h_{23} + \Delta h_{34} + \Delta h_{41} = \{Q_{12} + W_{12}\} + \{Q_{23} + W_{23}\}$$
$$+ \{Q_{34} + W_{34}\} + \{Q_{41} + W_{41}\} = 0$$

$$\sum \Delta h = \sum Q + \sum W = 0.$$

Therefore, the sum of heat and work transfer in a cycle is zero.

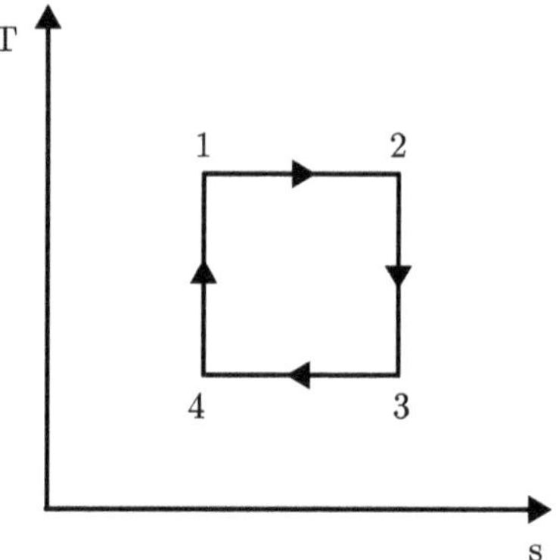

Figure 6.1. A property diagram of a simple four process forming a thermodynamic cycle.

The second law of thermodynamics introduces the concept of entropy, stating that for a reversible cycle,

$$0 = \sum \frac{\dot{Q}}{T} + s_{\mathrm{irr}}.$$

These principles are used to analyse engines, coolers, and heat pumps, each of which involves specific processes of heat absorption, heat rejection, and work input/output.

The performance of a thermodynamic cycle is expressed as the ratio of the desired output to the input effort:

$$\text{performance merit} = \frac{\text{desired}}{\text{effort}}.$$

6.6.1 Engine performance

An engine converts heat absorbed from a high-temperature reservoir (Q_h) into work (W) while rejecting a portion of the heat to a low-temperature reservoir (Q_c).

Figure 6.2 presents a thermodynamic diagram illustrating the global energy flows within a system undergoing a thermodynamic cycle. Unlike the idealized piston–gas systems studied earlier, where the working substance is assumed to undergo the same process at all points, real systems used for energy conversion involve more complex behaviors. In practical systems, heat input (Q_h) and heat output (Q_c) occur at specific locations within the system, corresponding to the hot region and cold region of the working substance. These regions are distinct and physically localized, and they drive the thermodynamic cycle by maintaining a temperature difference within the system.

It is crucial to note that the heat absorption and rejection do not take place uniformly across the system, but at these specific regions. The 'Hot' and 'Cold' labels in the figure represent the temperatures of the working substance in the hot and cold regions of the system, not abstract reservoirs. In the literature, many authors derive performance expressions assuming interactions between the engine and external hot

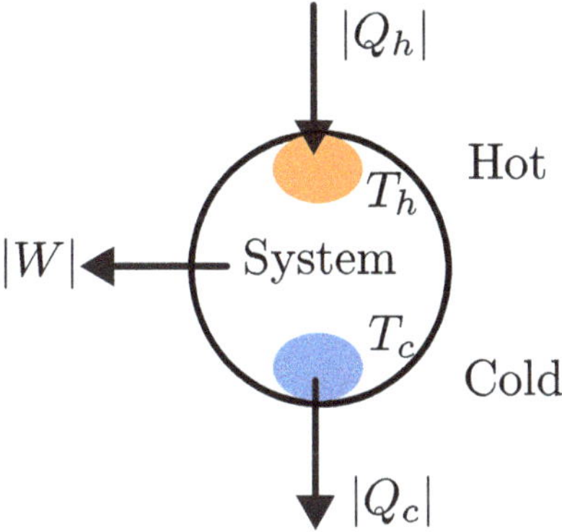

Figure 6.2. Schematic representation of energy flows in a heat engine cycle. The system absorbs heat $|Q_h|$ at a high internal temperature $|T_h|$, rejects heat $|Q_c|$ at a lower internal temperature $|T_c|$, and produces net work output $|W|$. The labels 'Hot' and 'Cold' refer to the relative temperatures within the system during different parts of the cycle.

and cold reservoirs. However, this approach is fundamentally flawed because reservoir temperatures are not system properties. Instead, the performance of the cycle depends on the temperatures of the working substance at these specific locations within the system.

The efficiency (η) of the engine is

$$\text{Efficiency,} \quad \eta = \frac{\text{work output (J or W)}}{\text{heat input (Jo r W)}} = \frac{|W|}{|Q_h|}.$$

Entropy changes during the heat transfer processes balance out over a cycle. For an engine,

$$\frac{|Q_h|}{T_h} - \frac{|Q_c|}{T_c} + s_{\text{irr}} = 0.$$

We will derive the performance expression for an engine using the T–S diagram provided. This derivation assumes that the *irreversible entropy generation within the system is zero*, meaning that all entropy changes in the system arise solely due to *heat transfer* between the working substance and its surroundings. This does not imply that the cycle is reversible, as heat transfer processes are inherently irreversible. However, we focus on systems where the internal processes (such as friction, turbulence, or mixing) do not contribute additional entropy generation. The entropy changes associated with heat transfer are explicitly accounted for, and these form the basis of the derivation.

6.6.1.1 Heat transfers in the cycle

1. **Heat transfer at the hot side** (Q_h): Heat is absorbed by the working substance during the process $1 \rightarrow 2$, where the working substance is at a temperature T_h. The entropy increase during this process is

$$\Delta S_{12} = S_2 - S_1.$$

The heat absorbed is given by

$$Q_H = T_H \, \Delta S_{12} = T_H \, (S_2 - S_1).$$

2. **Heat transfer at the cold side** (Q_c): Heat is rejected by the working substance during the process $3 \rightarrow 4$, where the working substance is at a temperature T_c. The entropy decrease during this process is

$$\Delta S_{34} = S_4 - S_3.$$

The heat rejected is

$$Q_c = T_c \, \Delta S_{34} = T_c \, (S_4 - S_3).$$

6.6.1.2 Entropy balance in the cycle

While heat transfer processes are irreversible, the assumption here is that no additional entropy is generated within the system beyond what is transferred during heat exchange. Thus, the total entropy change in the cycle is zero:

$$\sum \Delta S = \Delta S_{12} + \Delta S_{23} + \Delta S_{34} + \Delta S_{41} = 0.$$

In the cycle:
- $\Delta S_{23} = 0$ (isentropic process),
- $\Delta S_{41} = 0$ (isentropic process).

This simplifies the entropy balance to

$$\Delta S_{12} + \Delta S_{34} = 0.$$

Substituting the expressions for entropy changes during heat transfer:

$$\frac{Q_h}{T_h} + \frac{Q_c}{T_c} = 0 \ \text{ or } \ \frac{|Q_h|}{T_h} = \frac{|Q_c|}{T_c}.$$

6.6.1.3 Efficiency of the engine

From the first law of thermodynamics, the net work produced is

$$W = Q_h - Q_c.$$

Substituting this into the efficiency expression,

$$\eta = \frac{W}{Q_h} = \frac{Q_h - Q_c}{Q_h} = 1 - \frac{Q_c}{Q_h}.$$

Using the entropy relation $\left(\frac{|Q_h|}{T_h} = \frac{|Q_c|}{T_c} \right)$ the maximum efficiency of an engine is

$$\eta_{\max} = 1 - \frac{T_c}{T_h}.$$

6.6.1.4 Key insights and clarifications

1. **Zero internal irreversibility**: The assumption is that internal sources of entropy generation, such as friction or turbulence, are negligible. Hence, the only entropy changes are due to heat transfer. Therefore, the derived expression gives the maximum efficiency of an engine operating between two temperatures.
2. **Hot and cold regions**: The heat input (Q_h) occurs at a specific hot region with temperature T_h, and heat rejection (Q_c) occurs at a specific cold region with temperature T_c. These are *system properties* and not external reservoirs.
3. **Practical engine representation**: The cycle represents a realistic engine system with physically meaningful locations for heat input and output, ensuring that the derived efficiency is grounded in the system's actual behavior.

6.6.2 Cooler and heat pump performance

The analysis of a cooler or a heat pump follows a similar thermodynamic framework as the engine. However, the key difference lies in the direction of the cycle and the purpose of energy conversion. In a cooler or a heat pump, the cycle traverses in the *opposite direction* compared to an engine. Instead of converting heat into work, as in an engine, these devices *convert work* (or *electricity*) into thermal energy—either to extract heat from a cold region or to supply heat to a warm region (figure 6.3).

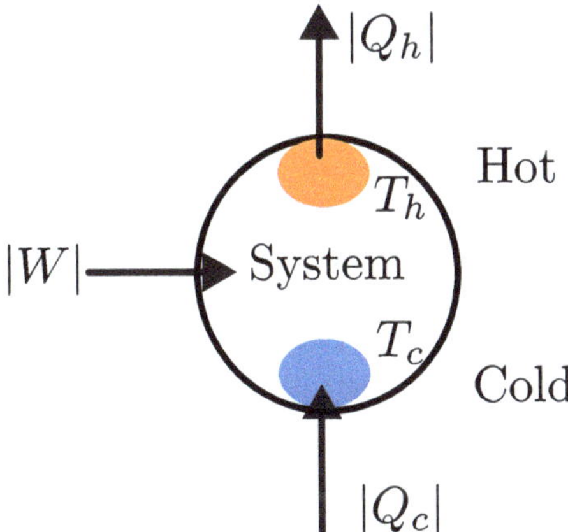

Figure 6.3. Schematic representation of a refrigeration or heat pump cycle. The system extracts heat Q_c at a lower internal temperature T_c, receives input work W, and rejects heat Q_h at a higher internal temperature T_h. The labels 'Cold' and 'Hot' refer to the system's internal temperatures during these interactions.

6.6.3 Key similarities and differences between a cooler and a heat pump

From a thermodynamic perspective, coolers and heat pumps operate on the *same principles* and share identical cycles. The distinction lies in the *desired energy output*:

1. **Cooler**:
   - The primary goal is *cooling*.
   - Heat (Q_c) is extracted from the cold region at temperature T_c. The rejected heat (Q_h) at temperature T_h is considered a by-product.
   - Performance is evaluated using the **coefficient of performance (COP)**, which measures the cooling effect relative to the work input.
2. **Heat pump**:
   - The primary goal is *heating*.
   - Heat (Q_h) supplied to the warm region at temperature T_h is the energy of interest. The extracted heat (Q_c) from the cold region is a means to achieve this.
   - Performance is also evaluated using the COP, but in this case, it measures the heating effect relative to the work input.

The cycle on a T–S diagram is reversed compared to an engine, reflecting the reverse flow of energy.

Coefficient of performance (COP)

The COP is a measure of the cycle's performance, expressed as the ratio of the desired energy output to the work input.

- **COP for a cooler:**

$$\text{COP}_{\text{cooler}} = \frac{|Q_c|}{|W|}.$$

  Here, Q_c is the desired cooling effect.
  The maximum COP of a cooler (without internal irreversibilities) is

$$\text{COP}_{\text{cooler,max}} = \frac{T_c}{T_h - T_c}.$$

- **COP for a heat pump:**

$$\text{COP}_{\text{heat pump}} = \frac{|Q_h|}{|W|}.$$

  Here, Q_h is the desired heating effect.
  The maximum COP of a heat pump (without internal irreversibilities) is

$$\text{COP}_{\text{heat pump,max}} = \frac{T_h}{T_h - T_c}.$$

6.7 Ideal versus real cycles

Thermodynamic cycles can be categorized as **ideal cycles** and **real cycles**, with ideal cycles serving as theoretical benchmarks for performance. While real cycles represent practical systems, ideal cycles provide insights into the maximum efficiency a system can achieve under perfect conditions.

The cycle discussed earlier in the T–S diagram is a **Carnot cycle**, characterized by $s_{\mathrm{irr}} = 0$. This means that the cycle is free of internal irreversibilities, and all entropy changes are solely due to heat transfer to or from the working substance. The Carnot cycle is the epitome of an ideal cycle, serving as the benchmark for all practical thermodynamic systems. However, it is important to note that the Carnot cycle is purely a thought experiment and cannot be realized in practice. Its assumptions, such as perfectly reversible processes and no heat transfer losses, are idealized and unattainable in real systems. Nonetheless, the Carnot cycle provides a theoretical upper limit for the performance of any thermodynamic system operating between two temperature limits.

6.7.1 Entropy generation and cycle irreversibility

The distinction between ideal and real cycles rests primarily on whether or not irreversible entropy generation is present. In an ideal cycle, all processes are quasistatic and reversible. Entropy is transferred through the system by heat exchange alone, and the net entropy generation within the system is zero. As a result, such cycles achieve maximum theoretical performance. However, these are idealizations.

In practical, real-world systems, some steps in the cycle inevitably deviate from reversibility. This deviation may be due to friction, rapid (non-quasistatic) compression or expansion, turbulent mixing, viscous flow, or uncontrolled heat transfer across finite temperature differences. All these effects lead to irreversible entropy generation within the system.

To visualize the consequences, consider the schematic in figure 6.4. Both panels represent cooling cycles, where the objective is to extract heat $\left| Q_c \right|$ at a low temperature T_c and reject it at a higher temperature T_h, supported by mechanical work $\left| W \right|$. In the left panel, the processes are quasistatic: there is no internal entropy generation, and all entropy delivered to the hot side originates from the cold-side heat absorption. Since entropy is a state function, the system's entropy returns to its initial value after a complete cycle. Thus, in this ideal case, the entropy flow out of the system (via heat at T_h) equals the entropy flow into it (via heat at T_c).

In contrast, the right panel shows a more realistic situation where some steps are non-quasistatic. While the system still performs the same cooling function, the presence of irreversible entropy generation leads to an excess of entropy that must be discharged to the hot side. This results in two important consequences. First, for the same heat absorption $\left| Q_c \right|$ at the cold end, the heat rejected $\left| Q_h \right|$ to the hot side is now larger. Second, more work input $\left| W \right|$ is required to drive the cycle. The internal

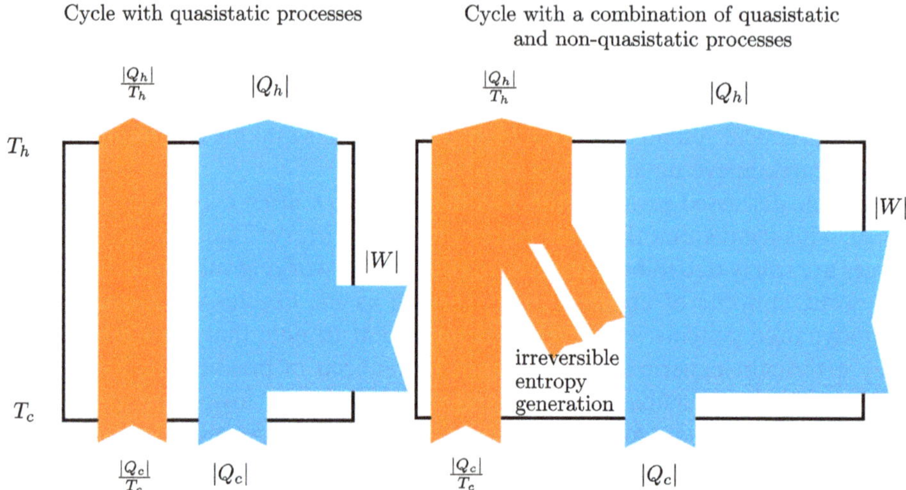

Figure 6.4. Comparison between a quasistatic and a non-quasistatic cooling cycle. In both cases, the system absorbs heat $|Q_c|$ at a lower temperature T_c, receives work input $|W|$, and rejects heat $|Q_h|$ at a higher temperature T_h. In the quasistatic case (left), all entropy transfer occurs via heat exchange, with no internal generation. In the non-quasistatic case (right), irreversible processes within the system generate additional entropy, increasing both the rejected heat and the required work input.

generation of entropy effectively degrades the performance, either by increasing energy input requirements or reducing the net cooling achieved.

The key message here is that the total entropy change of the system over one complete cycle is always zero, regardless of whether the cycle is reversible or not. But the entropy exported to the environment differs significantly between reversible and irreversible cycles, and this directly impacts the efficiency of real devices. Readers are encouraged to sketch a similar energy–entropy flow diagram for a heat engine cycle, identifying the entropy exchange and possible irreversible contributions in that case.

Key characteristics of ideal cycles

1. **No irreversible entropy generation**: Internal irreversibilities, such as friction or turbulence, are absent. The entropy changes observed are due only to heat transfer.
2. **Quasistatic processes**: All processes occur infinitely slowly to maintain equilibrium at every stage.
3. **Perfect heat transfer**: Heat transfer occurs only at the specific hot and cold regions in the system.
4. **Idealized working substance**: The working substance behaves ideally, following simple equations of state without deviations at extreme conditions.

6.8 Classification of cooling cycles: space-separated versus time-separated

The concept of refrigeration is not limited to a specific working fluid or mechanism. At its core, any system that can cyclically absorb heat at a low temperature and reject it at a higher temperature—while undergoing changes in entropy—qualifies as a refrigeration cycle. This universality allows the design of cooling systems based on gases, liquids, solids, or even fields (magnetic, electric). The essential requirement is that the working substance must undergo a thermodynamic cycle in which its entropy changes during interactions with heat reservoirs at different temperatures.

Cooling systems, particularly those operating on thermodynamic cycles, can be broadly classified into two categories based on how and where heat exchange occurs relative to the working material or system. This classification—**time-separated** versus **space-separated** cooling systems—is helpful in understanding the operational principles of different cooling technologies.

6.8.1 Time-separated cooling systems

In time-separated cooling systems, the processes of heat absorption (from the cold reservoir) and heat rejection (to the hot reservoir) occur at different times, often involving the same physical location or material. That is, the working material undergoes thermodynamic transformations in place, and at different stages in time, it is exposed alternately to cold and hot reservoirs to absorb or reject heat. These systems generally require intermittent thermal contact with the reservoirs, often realized through components like heat switches.

A good example of this class includes magnetocaloric, electrocaloric, and elastocaloric systems. In such systems, the working material exhibits a change in entropy and temperature under an applied field—magnetic, electric, or mechanical stress, respectively. For instance, in a magnetocaloric refrigerator:
- The material is first magnetized adiabatically, increasing its temperature.
- It is then thermally connected to a hot reservoir to reject heat.
- Afterward, the material is demagnetized, causing it to cool.
- It is then brought into contact with the cold reservoir to absorb heat.

This sequence repeats cyclically, but all heat exchange occurs at the same spatial location, while the timing and control of thermal contact (via switches or motion) define the cycle. The physical system itself does not move, but its thermodynamic state evolves over time in synchronization with the reservoir connections.

6.8.2 Space-separated cooling systems

In contrast, space-separated cooling systems involve movement of the working fluid or working material through physically distinct regions for heat absorption and heat rejection. The heat exchange occurs simultaneously but in different spatial locations —as the working substance flows through or oscillates between designated hot and cold zones.

A classical example is the vapor-compression refrigeration cycle (discussed in the next chapter), used in household refrigerators:
- The working fluid evaporates in the evaporator, located at the cold side, absorbing heat from the refrigerated space.

- It is then compressed and directed to the condenser, located in a different part of the system, where it rejects heat to the surroundings.
- The fluid cycles back after passing through an expansion valve, and the process repeats.

Another important example is the Stirling cryocooler, where the working gas oscillates between hot and cold ends of a cylinder. Each location is thermally anchored to a different reservoir, and the cycle is driven by pistons or displacers.

Thermoelectric devices, although solid-state, fall under the space-separated category. Here, electrical current flows through different regions of a thermoelectric element, causing heat to be absorbed at one junction and rejected at another. The two sides are spatially distinct—meaning no heat switch is required, and heat flows continuously while the device is powered.

6.9 Thermoelectric coolers: a non-gas-based cycle

The Peltier effect occurs when an electric current flows through a circuit made of two different materials (usually semiconductors). The current causes one junction to absorb heat (cooling) and the other junction to reject heat (heating). This heat transfer is a direct result of the charge carriers (electrons or holes) moving through the materials and carrying energy from one side of the system to the other.

Interestingly, the electrons in a thermoelectric circuit behave somewhat like a working fluid in traditional thermodynamic cycles. They are 'pumped' around the circuit, undergoing processes that involve energy absorption and rejection, akin to the heating and cooling processes of gases or liquids. However, instead of changes in pressure or volume, the electrons experience variations in energy states as they move through the different materials and across temperature gradients.

The basic process in a thermoelectric cooler involves the following steps:
1. **Heat absorption** (cooling side): Electric current passes through the thermoelectric module, causing heat to be absorbed from the cooler side (low temperature). This process creates a cooling effect at the interface, drawing heat away from the object being cooled.
2. **Heat rejection** (hot side): The heat absorbed from the cold side, along with the energy input from the electrical work, is rejected to the surroundings at the hot side (high temperature).
3. **Work input** (electric current): The energy required to move heat from the cold side to the hot side is provided by the electric current passing through the system.

This analogy of electrons acting as the 'working fluid' provides a bridge between the behavior of thermoelectric coolers and traditional thermodynamic systems. It emphasizes how energy transfer and work interaction are universal principles, even when no physical fluid is involved.

6.10 External temperatures and maximum performance

Thermodynamic systems, such as engines and coolers, interact with their surroundings to exchange heat and perform work. While we derived performance metrics in earlier sections based on the **internal system properties**—specifically the temperatures of the working substance at hot and cold regions—the reality is that these systems operate in conjunction with external surroundings. The temperatures of these surroundings, referred to as **external temperatures** ($T_{h,\text{ext}}$ and $T_{c,\text{ext}}$), directly influence the achievable performance of real-world systems.

Importantly, the external temperatures are typically *more extreme* than the internal system temperatures because of practical constraints, such as thermal resistance in heat exchangers and imperfect thermal interfaces. This section highlights how incorporating external temperatures into performance calculations results in a more realistic maximum performance that is lower than the ideal values based solely on system properties.

6.10.1 Engine performance with external temperatures

(Figure 6.5) In an engine, the working substance exchanges heat with two distinct regions in the system:

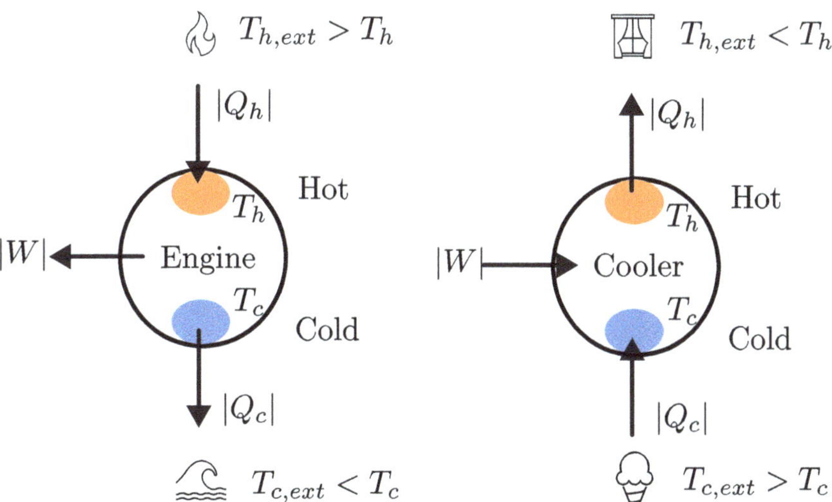

Figure 6.5. Schematic comparison of a heat engine (left) and a cooler or refrigerator (right), both interacting with external reservoirs. In the engine, heat is absorbed from a hot reservoir at $|T_h|$ and partially converted into work $|W|$, while the remaining heat $|Q_c|$ is rejected to a cold reservoir at $|T_c|$. In the cooler, work $|W|$ is supplied to extract heat $|Q_c|$ from a low-temperature reservoir and reject it as $|Q_h|$ to a higher-temperature sink. The external reservoirs ($T_{h,\text{ext}}$, $T_{c,\text{ext}}$) are shown to emphasize that they may differ from the system's internal temperatures $|T_h|$ and $|T_c|$.

- **Heat input** (Q_h): The working substance absorbs heat in the hot region at an internal temperature T_h. This heat originates from a high-temperature external source with a temperature $T_{h,\text{ext}}$.
- **Heat rejection** (Q_c): The working substance rejects heat in the cold region at an internal temperature T_c. This heat is transferred to a low-temperature external sink with a temperature $T_{c,\text{ext}}$.

Due to inefficiencies in heat exchangers and thermal resistances, the following relationships hold:

$$T_{h,\text{ext}} > T_h \text{ and } T_{c,\text{ext}} < T_c.$$

These more extreme external temperatures reduce the effective temperature difference driving the cycle. Consequently, when external temperatures are used to calculate the engine's efficiency, the maximum efficiency is lower than that derived from system properties alone. The realistic efficiency is expressed as

$$\eta_{\max} = 1 - \frac{T_{c,\text{ext}}}{T_{h,\text{ext}}}.$$

This expression represents the upper limit of performance for real engines when accounting for external conditions.

6.10.2 Cooler performance with external temperatures

(Figure 6.5) In a cooler, work is used to transfer heat from a low-temperature region to a higher-temperature region. The working substance undergoes thermodynamic processes in the system's cold and hot regions:

- **Heat absorption** (Q_c): The working substance absorbs heat in the cold region at an internal temperature T_c. This heat originates from the surroundings at a higher temperature $T_{c,\text{ext}}$.
- **Heat rejection** (Q_h): The working substance rejects heat in the hot region at an internal temperature T_h, transferring it to the surroundings at a lower temperature $T_{h,\text{ext}}$.

Due to similar inefficiencies as in engines, the external temperatures satisfy the following relationships:

$$T_{h,\text{ext}} < T_h \text{ and } T_{c,\text{ext}} > T_c.$$

The COP of a cooler, accounting for external temperatures, is given by

$$\text{COP}_{\max} = \frac{T_{c,\text{ext}}}{T_{h,\text{ext}} - T_{c,\text{ext}}}.$$

Note the difference between an engine and cooler.

6.11 Historical statements of the second law of thermodynamics

The second law of thermodynamics is often expressed through two classical formulations: the Clausius statement and the Kelvin–Planck statement. These statements describe fundamental limitations on heat and work interactions in cyclic processes. Although historically significant, these formulations are less emphasized in modern thermodynamics due to the deeper understanding provided by state variables such as entropy.

6.11.1 Clausius statement

The **Clausius statement** focuses on the behavior of heat transfer:
'It is impossible for a device operating in a cyclic process to transfer heat from a colder body to a hotter body without the input of external work.'

This principle explains the basic operation of refrigeration and heat pump systems. Heat naturally flows from high to low temperatures. To reverse this flow —moving heat from a cold region to a hot region—external work is required. For example:

- A refrigerator absorbs heat from its cold interior and rejects it to the warmer environment, powered by electrical work.
- A heat pump extracts heat from a colder outdoor space and transfers it to a warmer indoor area for heating, again requiring external energy.

The Clausius statement encapsulates the unidirectional nature of spontaneous heat transfer, emphasizing the necessity of work input to reverse this process.

6.11.2 Kelvin–Planck statement

The **Kelvin–Planck statement** focuses on the limits of energy conversion in heat engines:
'It is impossible to construct a device operating in a cyclic process that extracts heat from a single thermal reservoir and converts it entirely into work.'

This implies that no heat engine can have 100% efficiency. A portion of the heat absorbed from the hot region must always be rejected to a cold region. For instance:

- In a steam turbine, only part of the heat energy extracted from the high-temperature steam is converted to work; the rest is expelled as waste heat.

The Kelvin–Planck statement highlights the inherent inefficiency of heat engines, a direct consequence of the second law.

6.11.3 A modern perspective on these statements

While the Clausius and Kelvin–Planck statements were historically significant for understanding thermodynamic principles, modern thermodynamics, with its emphasis on *state variables* such as *entropy*, renders these formulations largely superfluous.

In this chapter, we derived the performance of thermodynamic cycles using the principle of entropy. For a complete cycle:

$$\Delta S_{\text{system}} = 0.$$

This ensures that the total entropy changes due to heat transfer and any entropy generated irreversibly must balance. From this single, fundamental principle:

- The unidirectional flow of heat described by the Clausius statement follows naturally, as entropy must increase when heat moves from hot to cold or external work must compensate otherwise.
- The Kelvin–Planck statement becomes self-evident because entropy changes impose a restriction on how much heat can be converted into work.

By focusing on entropy, modern thermodynamics provides a unified framework for understanding the second law without relying on separate, specialized statements.

6.12 Summary

This chapter provided an in-depth exploration of thermodynamic cycles, their mechanisms, and their practical implications. Key highlights include:

1. **Definition and purpose of cycles**: Thermodynamic cycles enable energy to be transformed repeatedly, making them indispensable in engines, power plants, refrigerators, and heat pumps.
2. **Ideal versus real cycles**: The Carnot cycle, representing the ideal benchmark for efficiency, was contrasted with real cycles, which include inefficiencies like friction and thermal losses.
3. **Analysis of key cycles**: Detailed analyses of the Brayton and Stirling cycles were presented. The Brayton cycle, used in gas turbines, operates with steady flow, while the Stirling cycle, common in cryogenic systems, involves oscillating flow and regenerator-based heat recycling.
4. **Performance metrics**: For engines, efficiency was discussed as the ratio of work output to heat input. For coolers and heat pumps, the coefficient of performance (COP) was introduced, emphasizing the desired cooling or heating effect relative to work input.
5. **Impact of external temperatures**: Practical performance limits were analysed by incorporating external temperatures into the efficiency and COP calculations, highlighting the importance of realistic assumptions in system design.

By connecting theoretical concepts with practical applications, this chapter lays a strong foundation for understanding and optimizing thermodynamic systems. It equips readers with the tools to evaluate the efficiency and performance of cycles in various engineering contexts.

IOP Publishing

A Classical Thermodynamics Toolkit

Srinivas Vanapalli

Chapter 7

Real gas properties and applications in thermodynamic cycles

7.1 Introduction

Thermodynamics extends far beyond idealized models, exploring the behavior of substances in realistic conditions where gases, liquids, and phase transitions present fascinating challenges. This chapter delves into the complexities of **real gas properties**, unveiling their relevance in critical industrial and scientific applications. From understanding deviations from ideal gas behavior to mastering the thermodynamic principles governing phase changes, this chapter lays the groundwork for tackling real-world engineering problems.

Key concepts such as **equations of state**, **phase diagrams**, and the generalization of thermodynamic laws to open systems are explored with a practical focus. Building on these foundations, we analyse essential thermodynamic cycles, such as the **vapor-compression refrigeration cycle** and the **Rankine cycle**, which form the backbone of modern cooling and power generation technologies. This chapter aims to equip readers with the tools to connect theory to practice, paving the way for solving challenges in fields such as refrigeration, energy systems, and cryogenics.

7.2 Limitations of the ideal gas model

The ideal gas law, $pV = nRT$, serves as a cornerstone in thermodynamics, providing a simplified relationship between pressure (p), volume (V), temperature (T), and the number of moles (n) of a gas. This model assumes that gas molecules have negligible volume and do not interact with one another, except during perfectly elastic collisions. While these assumptions are valid under many conditions, they break down in real-world applications where gases exhibit non-ideal behavior.

doi:10.1088/978-0-7503-6029-6ch7   7-1

7.2.1 When does the ideal gas model fail?

The ideal gas law works best at:
- **Low pressures**: Molecules are far apart, and their individual volumes are insignificant.
- **High temperatures**: The kinetic energy of molecules dominates any intermolecular forces.

However, at high pressures and low temperatures, real gases deviate significantly from ideal behavior. These deviations occur due to:
1. **Finite molecular volume**: As pressure increases, the volume of gas molecules becomes significant compared to the total volume of the gas.
2. **Intermolecular forces**: At low temperatures, attractive forces between molecules (e.g. van der Waals forces) cannot be ignored.

7.2.2 Why do we need real gas models?

The shortcomings of the ideal gas model necessitate more accurate representations of real gas behavior. These models account for:
- **Compressibility effects**, measured using the compressibility factor (Z).
- **Intermolecular attractions and repulsions**, captured through equations of state such as the van der Waals or Peng–Robinson equations.
- **Phase changes**, where gases transition to liquids or solids, requiring a completely different thermodynamic treatment.

By recognizing the limitations of the ideal gas law and adopting more comprehensive models, engineers and scientists can design systems that perform reliably under real-world conditions.

7.3 Equations of state for real gases

To address the limitations of the ideal gas law, more sophisticated models—referred to as equations of state—have been developed to describe the behavior of real gases. These equations incorporate corrections for molecular volume and intermolecular forces, providing a more accurate representation of gas behavior under non-ideal conditions.

7.3.1 The compressibility factor (Z)

One way to quantify deviations from ideal behavior is through the **compressibility factor**, defined as

$$Z = \frac{pV}{nRT}.$$

- For an ideal gas, $Z = 1$ at all conditions.
- For real gases, Z can deviate from unity, depending on pressure and temperature:

    ○ $Z < 1$: Indicates that attractive forces dominate, causing the gas to compress more than predicted by the ideal gas law.

    ○ $Z > 1$: Indicates that repulsive forces dominate, making the gas less compressible.

7.3.2 Common equations of state

Several equations of state have been developed to model real gas behavior. The most widely used include:

1. **van der Waals equation**: The van der Waals equation introduces two correction factors to the ideal gas law:

$$\left(P + \frac{a}{V^2}\right)(V - b) = RT.$$

    • a: Accounts for attractive forces between molecules.

    • b: Accounts for the finite volume of gas molecules.

    This equation is a good starting point for understanding real gas behavior, although its accuracy is limited at extreme conditions.

2. **Redlich–Kwong equation**: An improvement over van der Waals, especially for gases at moderate temperatures:

$$P = \frac{RT}{V - b} - \frac{a}{T^{0.5}V(V + b)}.$$

3. **Peng–Robinson equation**: Widely used in engineering, particularly in the chemical and petroleum industries, this equation provides a more accurate representation of gas–liquid equilibria:

$$P = \frac{RT}{V - b} - \frac{a\alpha(T)}{T^{0.5}V(V + b)}.$$

    • $\alpha(T)$: A temperature-dependent correction factor.

4. **Ideal versus real behavior at critical and saturation states**: Real gas equations are essential near the critical point, where small changes in temperature or pressure lead to significant density changes. They also help describe gas–liquid transitions during condensation or boiling.

7.3.3 Application of equations of state

Equations of state are vital for:

• **Designing compressors and turbines**: Ensuring accurate predictions of pressure, temperature, and volume changes.

• **Refrigeration systems**: Understanding the behavior of refrigerants under varying conditions.

• **Chemical processes**: Modeling reactions and separations where gases deviate significantly from ideal behavior.

7.4 Pressure–enthalpy (*p–h*) diagram

The *p–h* diagram is a powerful tool for visualizing the thermodynamic properties of substances across different phases. Figure 7.1 shows the *p–h* diagram for water illustrating the property landscape, providing insights into phase change behavior (figure 7.1).

7.4.1 Key features of the *p–h* diagram for water

1. **The two-phase dome**:
   - The dome-shaped region represents the liquid–vapor equilibrium. Within this dome, water coexists as a mixture of liquid and vapor.
   - The left boundary of the dome marks the saturated liquid line, where the first vapor bubbles form.
   - The right boundary marks the saturated vapor line, where the last liquid droplets evaporate.
2. **Critical point**:
   - At the top of the dome lies the critical point. Beyond this point, the liquid and vapor phases are indistinguishable, and the substance exhibits properties of both phases (supercritical fluid).
3. **Isotherms and phase transitions**:
   - Isotherms (constant temperature lines) show unique behaviors:
     - In the two-phase region, *p–h* isothermal lines are horizontal, representing constant pressure during phase change.
     - Outside the two-phase dome, the isotherms slope steeply, indicating single-phase (liquid or vapor) behavior.

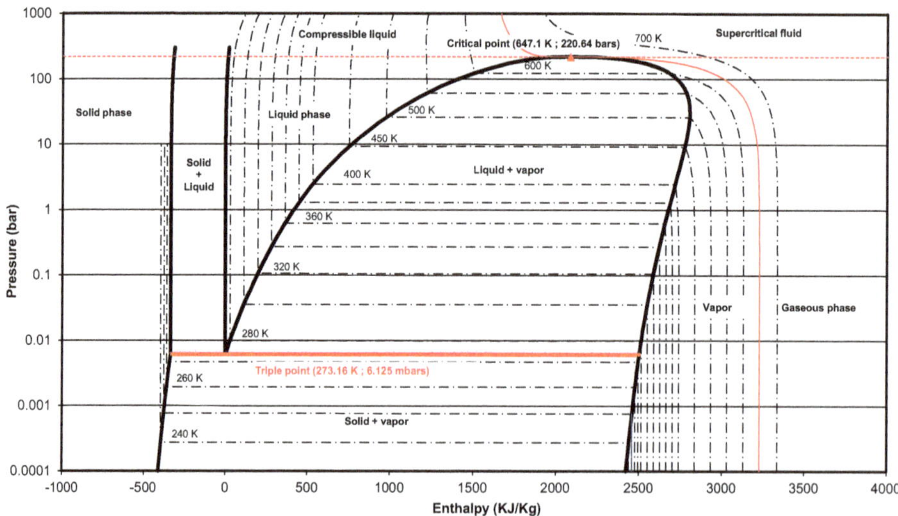

Figure 7.1. Pressure–enthalpy (*p–h*) diagram for water showing key thermodynamic regions, including subcooled liquid, two-phase mixture (liquid + vapor), superheated vapor, and supercritical fluid. The diagram also highlights the triple point (273.16 K; 6.125 mbar) and the critical point (647.1 K; 220.64 bar), along with selected isotherms and phase boundaries.

7.5 Pressure–temperature (*p–T*) diagram

The *p–T* diagram, often called the phase diagram, provides a comprehensive view of phase equilibrium. Figure 7.2 illustrates the *p–T* diagram for water, identifying key phase boundaries and points.

7.5.1 Key features of the *p–T* diagram for water

1. **Phase boundaries**:
   - The diagram delineates three main phase boundaries:
     - ○ **Sublimation line**: Separates the solid and vapor phases.
     - ○ **Melting line**: Separates the solid and liquid phases.
     - ○ **Boiling/condensation line**: Separates the liquid and vapor phases.
   - Each line represents equilibrium conditions where two phases coexist.
2. **Critical point**:
   - The critical point marks the upper limit of the boiling/condensation line, beyond which distinct liquid and vapor phases cease to exist.
3. **Triple point**:
   - The triple point is a unique state where solid, liquid, and vapor phases coexist in equilibrium. For water, this occurs at a pressure of 6.125 mbar and a temperature of 273.16 K.
4. **Solid region**:
   - The solid phase occupies a unique region in the *p–T* diagram. Depending on the temperature and pressure, water can form different solid structures, such as ice.

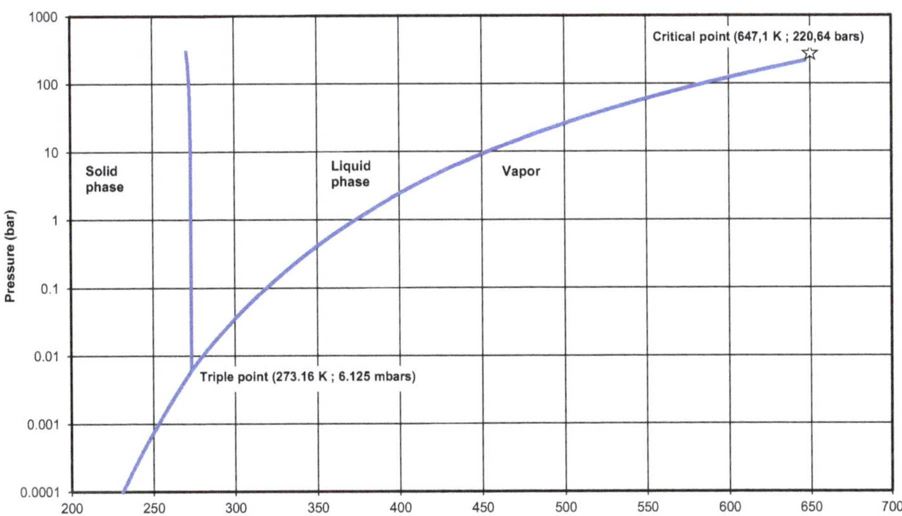

Figure 7.2. Pressure–temperature (*p–T*) diagram for water, illustrating the phase boundaries between solid, liquid, and vapor phases. The diagram highlights the triple point (273.16 K; 6.125 mbar) and the critical point (647.1 K; 220.64 bar), which represent key thermodynamic limits. The bold lines indicate equilibrium curves for phase transitions: sublimation, melting, and boiling/condensation.

7.6 Observations and real gas behavior

7.6.1 Key observations from p–T and p–h diagrams

1. **Real gas effects**:
   - In the supercritical region, water behaves neither as a typical liquid nor as a gas. The substance exhibits unique density and viscosity properties, useful in applications such as supercritical fluid extraction.
2. **Terminology**:
   - The term 'vapor' refers to a phase that can condense into a liquid under certain conditions. In contrast, the term 'gas' is often used to describe a phase that can be approximated by the ideal gas law.
3. **Application of phase diagrams**:
   - Engineers and scientists use these diagrams to design compressors, turbines, and heat exchangers that operate across single-phase and multi-phase regions.

The p–T and p–h diagrams are essential tools for understanding the thermodynamic properties of substances and analysing practical systems where real gas behavior and phase changes are significant.

7.6.2 A brief note on the Clapeyron equation

Phase change processes, such as boiling, condensation, melting, and sublimation, involve equilibrium between two phases of a substance. The conditions for this equilibrium are governed by the **Clapeyron equation**, which relates the slope of the phase boundary in a p–T diagram to the properties of the two coexisting phases:

$$\frac{dp}{dT} = \frac{\Delta h}{T \Delta v},$$

where:
- Δh is the enthalpy change (e.g. latent heat of vaporization or fusion),
- Δv is the specific volume difference between the two phases, and
- T is the absolute temperature.

This equation is invaluable for understanding and predicting the behavior of substances during phase changes. While the equation is introduced here conceptually, its derivation and applications will be discussed in detail in the next chapter.

7.7 Generalization of the first and second laws to an open system

Thermodynamic systems often involve the flow of mass across their boundaries. Such systems, referred to as **open systems**, include components such as compressors, turbines, and heat exchangers. To analyse these systems, the first and second laws of thermodynamics must be adapted to account for the transport of energy and entropy with the entering and exiting mass.

7.7.1 First law of thermodynamics for an open system

The first law of thermodynamics states that energy cannot be created or destroyed. For an open system with mass flow, heat transfer ($\dot{Q}$), work transfer ($\dot{W}$), and energy transport by mass ($\dot{m}h$) must be considered. The general form of the first law is

$$\dot{Q} + \dot{W} + \sum \dot{m}_{\text{in}} h_{\text{in}} = \sum \dot{m}_{\text{out}} h_{\text{out}} + \frac{dE}{dt},$$

where:
- h is the specific enthalpy ($h = u + pv$),
- $\dot{m}_{\text{in}}$ and $\dot{m}_{\text{out}}$ are the mass flow rates into and out of the system, and
- dE/dt is the rate of change of the system's total energy.

In **steady-state conditions**, the system's energy does not change over time ($\frac{dE}{dt} = 0$), and the equation simplifies to

$$\dot{Q} + \dot{W} + \dot{m}_{\text{in}} h_{\text{in}} = \dot{m}_{\text{out}} h_{\text{out}}.$$

This form is commonly used for analysing flow devices such as turbines and compressors.

7.7.2 Second law of thermodynamics for an open system

The second law introduces entropy (S) as a measure of system disorder. For an open system, the general form of the second law is

$$\frac{\dot{Q}}{T} + \dot{m}_{\text{in}} s_{\text{in}} - \dot{m}_{\text{out}} s_{\text{out}} + \dot{s}_{\text{gen}} = \frac{dS}{dt},$$

where:
- s is the specific entropy,
- $\dot{s}_{\text{gen}}$ represents entropy generation due to irreversibilities.

In **steady-state conditions**, entropy does not accumulate in the system ($dS/dt = 0$), reducing the equation to

$$\frac{\dot{Q}}{T} + \dot{m}_{\text{in}} s_{\text{in}} + \dot{s}_{\text{gen}} = \dot{m}_{\text{out}} s_{\text{out}}.$$

7.8 Application of the laws

The generalized first and second laws are applied to analyse key components of thermodynamic systems. Each component plays a crucial role in energy conversion or transfer, and their analysis relies on the properties of the substance.

7.8.1 Heat exchangers

Heat exchangers transfer energy between two fluid streams without direct mixing. Consider the example in figure 7.3, where water enters the heat exchanger as liquid at 290 K and exits as saturated steam at 1 bar.

Using the first law of thermodynamics:

$$\dot{Q} = \dot{m}(h_2 - h_1).$$

From the steam tables:
- Enthalpy of liquid water (h_1) at 290 K = 70.8 J g^{-1}.
- Enthalpy of saturated steam (h_2) at 1 bar = 2674.9 J g^{-1}.

For a mass flow rate of 1 g s^{-1},

$$\dot{Q} = 1 \text{ g s}^{-1} \times (2674.9 - 70.8) \text{ J g}^{-1} = 2604.1 \text{ W}.$$

This energy input transitions water from the subcooled liquid region into the two-phase dome, visible in the p–h diagram for water.

The entropy change during this process is

$$\dot{S} = \dot{m}(s_2 - s_1),$$

where:
- Entropy at the initial state (s_1) = 0.2513 J g^{-1} · K^{-1}.
- Entropy at the final state (s_2) = 7.3588 J g^{-1} · K^{-1}.

For the same mass flow rate:

$$\dot{S} = 1 \text{ g s}^{-1} \times (7.3588 - 0.2513) \text{ J g}^{-1} \text{ K}^{-1} = 7.1075 \text{ W K}^{-1}.$$

This example demonstrates how heat exchangers operate within the p–h property space and how energy and entropy are accounted for using the first and second laws.

7.8.2 Throttling devices

Throttling is an isenthalpic process, meaning that the enthalpy of the fluid remains constant across the throttling device. Since no heat (Q = 0) or work (W = 0) is exchanged with the surroundings, and changes in kinetic and potential energy are

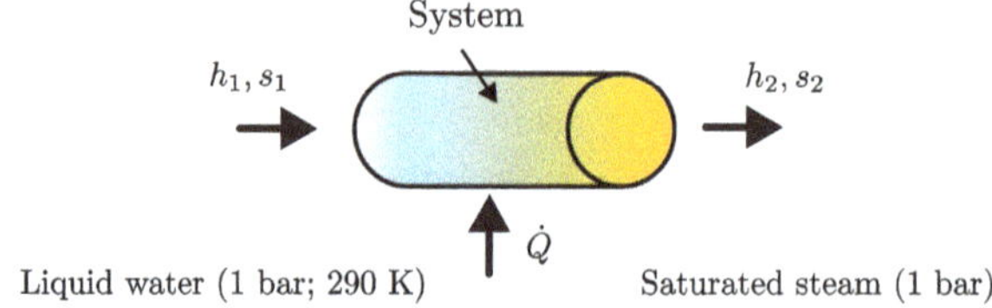

Figure 7.3. Schematic representation of a heat exchanger where liquid water at 1 bar and 290 K enters the system and is converted into saturated steam at the same pressure.

negligible, the first law of thermodynamics simplifies to $h_{\text{in}} = h_{\text{out}}$. Figure 7.4 shows examples of throttling devices.

Entropy increases in throttling due to irreversibilities. From the second law,

$$\dot{S}_e = \dot{S}_i + \dot{S}_{\text{gen}}.$$

The p–h diagrams clearly illustrate the temperature and entropy changes, even as enthalpy remains constant.

7.8.3 Compressors and pumps

Compressors and pumps are designed to increase the pressure of gases and liquids (figure 7.5). In the p–h diagram for water:
- A pump moves a liquid from a low-pressure state (subcooled region) to a higher-pressure state.
- A compressor performs a similar function for gases, but the process involves significant temperature and entropy changes.

Using the first law of thermodynamics:

$$\dot{Q} + \dot{W} = \dot{m}(h_o - h_i).$$

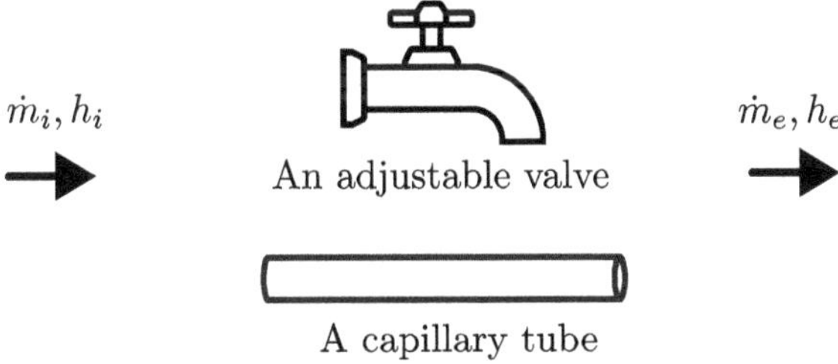

Figure 7.4. Examples of throttling devices: an adjustable expansion valve and a capillary tube. In both cases, the mass flow rate and enthalpy remain constant across the device. These components operate without heat or work interaction with the surroundings, and the process is characterized by a significant drop in pressure and an increase in entropy due to irreversibilities.

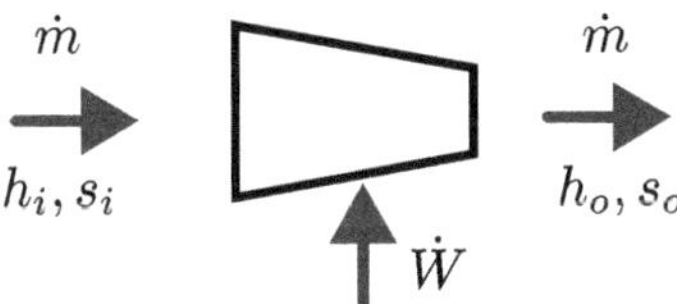

Figure 7.5. Schematic of a compressor or pump, where a working fluid is pressurized by mechanical work input. The process typically results in increased enthalpy and, in real systems, also increased entropy due to irreversibilities.

If the process deviates from ideality, additional entropy generation occurs, reducing efficiency.

$$\dot{m}\dot{S}_o = \dot{m}\dot{S}_i + \dot{S}_{\text{gen}}.$$

7.9 Vapor-compression cycle

The vapor-compression cycle is the cornerstone of mechanical refrigeration systems, widely used in the food and pharmaceutical industries for freezing and cooling applications. It relies on a closed-loop cycle where the working fluid, called the refrigerant, undergoes thermodynamic processes to transfer heat from a low-temperature region (cooling) to a high-temperature region (heat rejection).

In this cycle, the refrigerant is the system, and its state evolves through four main processes: throttling, evaporation, compression, and condensation. These processes are best visualized on a pressure–enthalpy (p–h) diagram, as shown in figure 7.6. The thermodynamic properties of the refrigerant at the key states are summarized in table 7.1.

7.9.1 Throttling process (1 → 2)

- *Description*: The refrigerant at state 1 (high-pressure, saturated liquid) passes through an expansion valve. This is an irreversible process during which no heat ($Q = 0$) or work ($W = 0$) is exchanged with the surroundings.

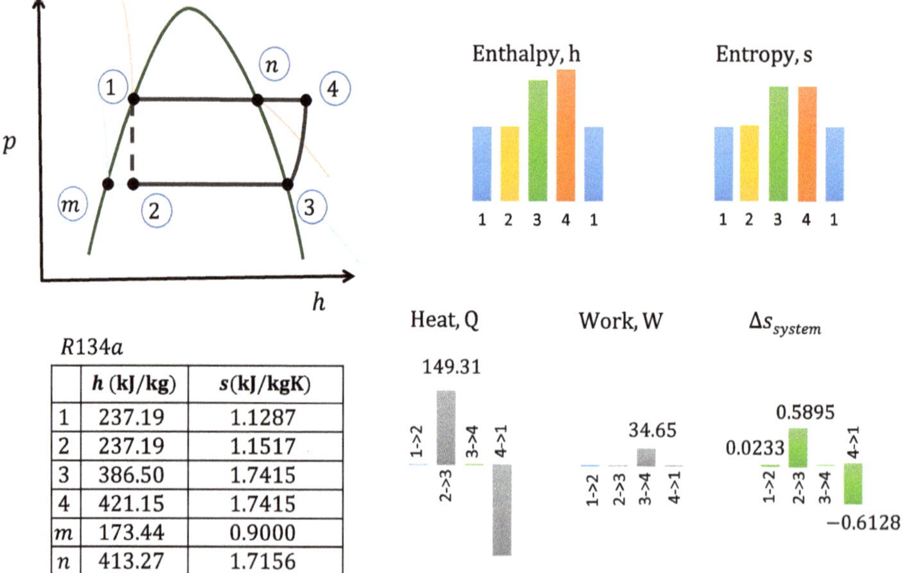

R134a	h (kJ/kg)	s (kJ/kgK)
1	237.19	1.1287
2	237.19	1.1517
3	386.50	1.7415
4	421.15	1.7415
m	173.44	0.9000
n	413.27	1.7156

Figure 7.6. Thermodynamic analysis of a vapor-compression refrigeration cycle using R134a, shown on a pressure–enthalpy (p–h) diagram. The cycle consists of four main processes: throttling (1 → 2), evaporation (2 → 3), compression (3 → 4), and condensation (4 → 1). Bar charts illustrate the enthalpy and entropy values at each state. The accompanying table provides thermodynamic properties at key points, including intermediate states m and n. The plots for heat (Q), work (W), and system entropy change (Δ_{system}) quantify energy and entropy balances across each process.

Table 7.1. Thermodynamic properties of R134a at four key states in the vapor-compression refrigeration cycle.

State	p (bar)	T (K)	h (kJ kg^{-1})	s (kJ kg^{-1}K^{-1})
1	**7.0**	**300.00**	237.19	1.1287
2	**1.3**	**253.13**	237.19	1.1517
3	1.3	253.13	386.50	1.7415
4	7.0	307.68	421.15	1.7415

Table 7.2. Energy and entropy balances for each process in the vapor-compression refrigeration cycle using R134a. The table reports changes in enthalpy, entropy, heat transfer (Q), and work (W) per unit mass for each segment of the cycle. The total enthalpy and entropy changes over the full cycle are zero, confirming the cyclic nature of the system.

Process	Δh (kJ kg^{-1})	Δs (kJ kg^{-1} K^{-1})	Q (kJ kg^{-1})	W (kJ kg^{-1})
1–2	0	0.0233	0	0
2–3	149.31	0.5895	149.31	0
3–4	34.65	0	0	34.65
4–1	−183.96	−0.6128	−183.96	0
Cycle	0	**0**	−34.65	34.65

- *Energy analysis*: From the first law of thermodynamics,

$$Q + W = h_2 - h_1.$$

  Since $Q = 0$ and $W = 0$,

$$h_2 = h_1.$$

  The enthalpy remains constant ($h_1 = h_2 = 237.19\,\text{kJ kg}^{-1}$). However, the refrigerant undergoes a significant drop in pressure, entering the two-phase region.
- *Entropy change*:

  During throttling, the entropy of the refrigerant increases ($s_2 > s_1$), even though no heat transfer occurs. This increase is due to irreversibilities inherent in the throttling process. As a result, the cycle cannot achieve thermodynamic maximum performance, as discussed in earlier chapters.

7.9.2 Evaporation process (2 → 3)

- *Description*: The refrigerant, now a low-pressure liquid–vapor mixture, enters the evaporator. It absorbs heat ($Q > 0$) from the surroundings, causing the liquid to vaporize completely into a saturated vapor at state 3. This is the cooling process.

- *Energy analysis*: From the first law,

$$Q + W = h_3 - h_2.$$

Since $W = 0$ (no work input/output),

$$Q = h_3 - h_2 = 386.50 - 237.19 = 149.31 \text{ kJ kg}^{-1}.$$

- *Entropy change*:
  The entropy of the refrigerant increases significantly during evaporation ($s_3 > s_2$) due to heat absorption. This can be observed in the bar chart for entropy in figure 7.6.

7.9.3 Compression process (3 → 4)

- *Description*: The refrigerant, now a low-pressure saturated vapor, is compressed adiabatically (isentropically) by a compressor to a high-pressure superheated vapor. This process requires work input ($W > 0$).
- *Energy analysis*: From the first law,

$$Q + W = h_4 - h_3.$$

Since $Q = 0$ (adiabatic process),

$$W = h_4 - h_3 = 421.15 - 386.50 = 34.65 \text{ kJ kg}^{-1}.$$

- *Entropy change*:
  For an ideal isentropic compression, the entropy remains constant ($s_4 = s_3$), as shown in the entropy bar chart. However, in real systems, some irreversibilities may slightly increase entropy.

7.9.4 Condensation process (4 → 1)

- *Description*: The high-pressure, superheated vapor enters the condenser, where it rejects heat ($Q < 0$) to the surroundings. The refrigerant transitions from a superheated vapor at state 4 to a high-pressure saturated liquid at state 1, completing the cycle.
- *Energy analysis*: From the first law,

$$Q + W = h_1 - h_4.$$

Since $W = 0$ (no work input/output),

$$Q = h_1 - h_4 = 237.19 - 421.15 = -183.96 \text{ kJ kg}^{-1}.$$

The negative sign indicates heat rejection.
- *Entropy change*:
  During condensation, the entropy of the refrigerant decreases ($s_1 < s_4$) as heat is removed. This is consistent with the entropy bar chart. The energy and entropy balances for each process are listed in table 7.2.

7.9.5 Entropy and enthalpy analysis of the cycle

The vapor-compression cycle highlights critical insights about the thermodynamic behavior of the refrigerant.

1. **System entropy**: The net entropy change of the refrigerant over a complete cycle is zero,

$$\Delta s_{\text{system}} = 0.$$

   This is because the refrigerant returns to its initial thermodynamic state after completing one cycle. However, entropy generation due to irreversibilities ($s_{\text{irr}} > 0$) occurs, particularly during the throttling process ($1 \rightarrow 2$), where heat transfer is zero, but the entropy of the refrigerant increases.

2. **System enthalpy**: Similarly, the enthalpy of the refrigerant remains unchanged after completing the cycle,

$$\Delta h_{\text{system}} = 0.$$

   Since the cycle is closed and the refrigerant returns to its initial state (state 1), there is no net change in enthalpy over the cycle.

7.9.6 The effect of irreversibilities on cycle performance

While the ideal vapor-compression cycle assumes isentropic compression and no irreversibilities, real systems experience losses that impact performance. Let us analyse the consequences of inefficiencies in the compressor and other components, focusing on entropy generation, enthalpy changes, and their thermodynamic implications.

7.9.6.1 Inefficient compression and its consequences

In an ideal compression process ($3 \rightarrow 4$), the refrigerant undergoes an **isentropic process**, meaning there is no entropy change ($s_4 = s_3$). However, in a real compressor with inefficiencies, irreversibilities lead to an increase in entropy ($s_4 > s_3$) at the outlet. This entropy generation has direct consequences on the thermodynamic state at point 4.

Impact on enthalpy at state 4
- With increased entropy ($s_4 > s_3$) and the same pressure p_4 (since the compressor still achieves the same high pressure as in the ideal case), the temperature of the refrigerant at state 4 is higher than in the ideal case.
- Higher temperature means higher enthalpy at state 4 ($h_4^{\text{real}} > h_4^{\text{ideal}}$).

Impact on compressor work (W)
- The work done by the compressor is the enthalpy difference between states 4 and 3,

$$W = h_4 - h_3.$$

- Since $h_4^{\text{real}} > h_4^{\text{ideal}}$, the work input required for compression increases in the presence of irreversibilities.
- Importantly, states 2 and 3 remain unchanged because the throttling and evaporation processes are unaffected by the compressor inefficiency. This means the required cooling ($Q_c = h_3 - h_2$) does not change.

Impact on heat rejection (Q_h)
- The condenser must reject the extra energy introduced into the system due to the increased compressor work.
- Heat rejection is calculated as

$$Q_h = h_4 - h_1.$$

- Since $h_4^{\text{real}} > h_4^{\text{ideal}}$, the enthalpy difference between states 4 and 1 increases, leading to higher heat rejection.

7.9.6.2 Key observations on work and heat transfer
1. **Work input increases**:
   Any irreversibilities in the system increase the required work input for a given cooling effect (Q_c). This is because the entropy generated in the system must be balanced by increased heat rejection to maintain the cyclic operation.
2. **Heat rejection increases**:
   With higher work input (W) due to irreversibilities, the condenser must reject more heat to the surroundings. This increases the enthalpy difference between states 4 and 1 ($h_4 - h_1$).

7.9.6.3 Example: Inefficient compression
Let us consider an example with an inefficient compressor where the outlet entropy increases:
- **Ideal case**: $s_3 > s_4^{\text{ideal}}$, $h_4^{\text{ideal}} = 421.15$ kJ kg^{-1}.
- **Real case**: $s_4^{\text{real}} > s_4^{\text{ideal}}$, $h_4^{\text{real}} = 430.00$ kJ kg^{-1} (hypothetical).

In the real case:
- **Work input**:

$$W^{\text{real}} = h_4^{\text{real}} - h_3 = 430.00 - 386.50 = 43.50 \text{ kJ kg}^{-1}.$$

   Compared to the ideal case,

$$W^{\text{ideal}} = h_4^{\text{ideal}} - h_3 = 421.15 - 386.50 = 34.65 \text{ kJ kg}^{-1}.$$

   The compressor must now perform 8.85 kJ kg^{-1} more work for the same cooling effect.
- **Heat rejection**:

$$Q_h^{\text{real}} = h_4^{\text{real}} - h_1 = 430.00 - 273.19 = 192.81 \text{ kJ kg}^{-1}.$$

Compared to the ideal case,

$$Q_h^{\text{real}} = h_4^{\text{real}} - h_1 = 421.15 - 273.19 = 183.96 \text{ kJ kg}^{-1}.$$

Heat rejection increases by 8.85 kJ kg^{-1}, matching the increased work input.

Irreversibilities increase the total entropy of the system and surroundings, requiring additional work input to achieve the same cooling effect. Consequently, the **coefficient of performance (COP)** decreases because the same cooling effect now requires more work input:

$$\text{COP} = \frac{Q_c}{W} \left(\text{COP decreases as } W \text{ increases} \right).$$

7.9.7 Heat pumps and the vapor-compression refrigeration cycle

Although this chapter primarily describes the vapor-compression refrigeration cycle, the principles and thermodynamic analysis apply equally to heat pumps. The key distinction is in the intended benefit: while refrigeration focuses on extracting heat (Q_c) from a low-temperature region, heat pumps are designed to deliver heat (Q_h) to a high-temperature region as the useful output. The thermodynamic processes, including compression, condensation, throttling, and evaporation, remain identical, and the analysis using the first and second laws of thermodynamics is unchanged. The efficiency of a heat pump is measured by its COP for heating, given by

$$\text{COP}_{\text{heat}} = \frac{Q_h}{W}.$$

This highlights the flexibility of the vapor-compression cycle in serving both cooling and heating applications.

7.10 Rankine cycle

The Rankine cycle is a thermodynamic cycle commonly used in power plants to generate electricity from heat sources such as coal, natural gas, nuclear reactions, or solar energy. The working fluid, typically water, undergoes a series of processes, converting heat into mechanical work and completing a closed-loop cycle.

In a Rankine cycle, the working fluid transitions between liquid and vapor phases, making it a prime example of a two-phase thermodynamic cycle. The cycle consists of four main processes, which are shown on the pressure–volume ($p-V$) diagram provided in figure 7.7.

7.10.1 Main processes of the Rankine cycle

1. **Isentropic compression ($A \rightarrow B$):**
   - In this process, the liquid working fluid (water) is pumped from a low-pressure state (A) to a high-pressure state (B) using a pump.
   - The process is assumed to be adiabatic and isentropic, meaning no heat is transferred, and the entropy remains constant ($s_A = s_B$).
   - Work is done on the fluid by the pump: $W_{\text{pump}} = h_B - h_A$.

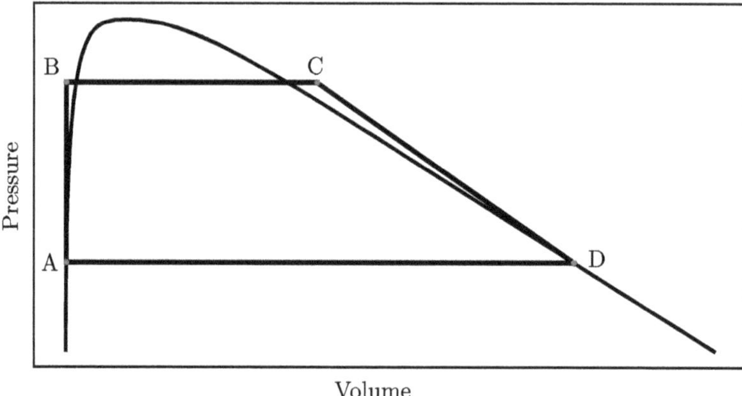

Figure 7.7. Pressure–volume (p–V) diagram of the Rankine cycle, illustrating the four main processes: isentropic compression ($A \rightarrow B$), isobaric heat addition ($B \rightarrow C$), isentropic expansion ($C \rightarrow D$), and isobaric heat rejection ($D \rightarrow A$). While the Rankine cycle is often represented in pressure–enthalpy (p–h) diagrams for engineering calculations, this figure uses the p–V format to emphasize that thermodynamic cycles can be visualized on different property planes to aid conceptual understanding.

2. **Isobaric heat addition ($B \rightarrow C$):**
   - The high-pressure liquid enters the boiler, where it absorbs heat (Q_{in}) from an external heat source.
   - The liquid is converted into high-pressure, high-temperature vapor (state C).
   - The enthalpy increase is: $Q_{\text{in}} = h_C - h_B$.
3. **Isentropic expansion ($C \rightarrow D$):**
   - The vapor expands through a turbine, producing work (W_{turbine}).
   - This process is also assumed to be adiabatic and isentropic ($s_C = s_D$).
   - The work *output* is $W_{\text{turbine}} = h_C - h_D$.
4. **Isobaric heat rejection ($D \rightarrow A$):**
   - The low-pressure vapor enters the condenser, where it rejects heat (Q_{out}) to the surroundings, condensing into a saturated liquid.
   - The enthalpy *decrease* is $Q_{\text{out}} = h_D - h_A$.

7.10.2 Worked example

- *Assumptions*:

  ○ Working fluid: water.
  ○ Pressure at state A: 0.1 bar (condenser pressure).
  ○ Pressure at state B: 30 bar (boiler pressure).
  ○ Turbine inlet temperature: 500 °C.
  ○ All processes are ideal (isentropic compression and expansion).

- *Using steam tables*:

  ○ At state A ($p_A = 0.1$ bar, saturated liquid),
     $h_A = 191.81$ kJ kg^{-1}, $s_A = 0.6492$ kJ kg^{-1} K^{-1}.
  ○ At state B ($p_B = 30$ bar, subcooled liquid),
     approximate using pump work,

  $$W_{\text{pump}} = v_A(p_B - p_A),$$

  with $v_A = 0.001043$ m^3 kg^{-1},

  $$W_{\text{pump}} = 0.001043 \times (30 \times 10^5 - 0.1 \times 10^5) = 3.13\,\text{kJ kg}^{-1}$$

  $$h_B = h_A + W_{\text{pump}} = 191.81 + 3.13 = 194.94\,\text{kJ kg}^{-1}.$$

  ○ At state C ($p_C = 30$ bar, $T_C = 500\,°\text{C}$),
     from steam tables,

  $$h_C = 3408.1\,\text{kJ kg}^{-1}, \quad s_C = 6.597\,\text{kJ kg}^{-1}\,\text{K}^{-1}.$$

  ○ At state D ($p_D = 0.1$ bar, isentropic expansion $s_D = s_C = 6.597$ kJ kg^{-1} K^{-1}),
     from steam tables,

  $$h_D = 2206.5\,\text{kJ kg}^{-1}.$$

7.10.3 Energy analysis

Pump work:

$$W_{\text{pump}} = 3.13\,\text{kJ kg}^{-1}.$$

Turbine work:

$$W_{\text{turbine}} = h_C - h_D = 3408.1 - 2206.5 = 1201.6\,\text{kJ kg}^{-1}.$$

Heat input:

$$Q_{\text{in}} = h_C - h_B = 3408.1 - 194.94 = 3213.2\,\text{kJ kg}^{-1}.$$

Heat rejected:

$$Q_{\text{out}} = h_D - h_A = 2206.5 - 191.81 = 2014.7\,\text{kJ kg}^{-1}.$$

7.10.4 Thermal efficiency

The thermal efficiency of the Rankine cycle is

$$\eta = \frac{W_{\text{net}}}{Q_{\text{in}}},$$

where

$$W_{\text{net}} = W_{\text{turbine}} - W_{\text{pump}} = 1201.6 - 3.13 = 1198.5 \text{ kJ kg}^{-1}$$

$$\eta = \frac{1198.5}{3213.2} = 0.373 = 37.3\%.$$

7.11 Summary

In this chapter, we examined the critical role of **real gas properties** in thermodynamics and their application to engineering systems. The highlights include:

1. **Real gas behavior**:
   - We began by exploring when and why real gases deviate from the ideal gas model due to finite molecular volume and intermolecular forces.
   - Common equations of state, such as van der Waals and Peng–Robinson, were introduced to accurately predict real gas behavior.

2. **Phase changes and thermodynamic analysis**:
   - Phase diagrams (p–T and p–h) were presented as powerful tools for understanding phase transitions.
   - The thermodynamic properties during boiling, condensation, and supercritical transitions were described, highlighting their practical implications.

3. **First and second laws for open systems**:
   - The adaptation of thermodynamic laws to open systems such as compressors, turbines, and heat exchangers was discussed.
   - Entropy generation and its impact on system performance were analysed in detail, showcasing their relevance in engineering design.

4. **Thermodynamic cycles**:
   - The **vapor-compression refrigeration cycle** was systematically explained, emphasizing the refrigerant as the system. Key processes, heat and work transfer, and entropy changes were analysed to understand its limitations and performance.
   - The **Rankine cycle**, a cornerstone of power generation, was introduced. A worked example illustrated how energy flows and thermal efficiency are calculated for a realistic system.

5. **Impact of irreversibilities**:
   - Irreversibilities, particularly in components such as compressors and throttling devices, were shown to degrade cycle performance.
   - Strategies for minimizing entropy generation to optimize efficiency were discussed.

IOP Publishing

A Classical Thermodynamics Toolkit

Srinivas Vanapalli

Chapter 8

Free energy and the spontaneity of a process

8.1 Introduction

This chapter provides an in-depth exploration of the fundamental thermodynamic concepts necessary for understanding the spontaneity of physical and chemical processes. Free energy, particularly Gibbs free energy, plays a pivotal role in predicting whether a process can occur spontaneously. This chapter integrates key thermodynamic principles, such as entropy, enthalpy, and work, to create a framework that allows for the systematic analysis of processes, from phase transitions to chemical reactions.

We begin by discussing the concept of spontaneity through the example of the melting of ice. This sets the stage for the introduction of Gibbs free energy and how it can be used to determine the spontaneity of various processes. Further sections expand on the application of Gibbs free energy in both physical and chemical processes, providing examples such as the combustion of hydrogen, fuel cell reactions, and electrolysis.

Additionally, the chapter introduces the Clausius–Clapeyron relation, a critical equation for understanding phase transitions, particularly in the context of vaporization and sublimation. The relation allows for the prediction of changes in vapor pressure with temperature and pressure, aiding in the understanding of phase equilibria.

The chapter concludes by discussing the Helmholtz free energy, which is used in scenarios where systems are at constant volume and temperature, in contrast to the Gibbs free energy's application in constant pressure and temperature systems. Through examples and derivations, this chapter equips readers with the tools to solve practical and theoretical thermodynamics problems in various domains such as material science, chemical engineering, and environmental science.

8.2 Spontaneity example: ice melting to water

To understand the concept of free energy and the spontaneity of a process, let us begin with a familiar example: the melting of ice into water. On Earth, at a temperature of 20 °C, this process occurs naturally and spontaneously. However, if we consider the same process on Mars, where the ambient temperature is well below 0 °C, the outcome is very different. By analysing these cases, we can develop a systematic understanding of the conditions for spontaneity and introduce the concept of Gibbs free energy.

8.2.1 The role of entropy in spontaneity

The total entropy change (ΔS_{total}) for a process can be expressed as the sum of the entropy changes of the system (ΔS_{sys}) and the surroundings (ΔS_{surr}):

$$\Delta S_{\text{total}} = \Delta S_{\text{sys}} + \Delta S_{\text{surr}}.$$

For processes occurring at constant pressure and temperature, the entropy change of the surroundings is directly related to the heat exchanged with the system (Q),

$$\Delta S_{\text{surr}} = -\frac{Q}{T_{\text{surr}}},$$

where T_{surr} is the temperature of the surroundings, and Q is the heat absorbed or released by the system at constant pressure. Using the enthalpy change of the system ($\Delta H_{\text{sys}} = Q$),

$$\Delta S_{\text{surr}} = -\frac{\Delta H_{\text{sys}}}{T_{\text{surr}}}.$$

Substituting this relationship into the expression for ΔS_{total}, we obtain

$$\Delta S_{\text{total}} = \Delta S_{\text{sys}} - \frac{\Delta H_{\text{sys}}}{T_{\text{surr}}}.$$

A process is considered **spontaneous** if the total entropy change is positive, i.e.

$$\Delta S_{\text{total}} > 0.$$

8.2.2 Example: Ice melting on Earth and Mars

Let us evaluate the spontaneity of ice melting on Earth (20 °C) and Mars (−50 °C). Assume the following thermodynamic values for water:

$$\Delta H_{\text{fus}} = 334 \text{ kJ kg}^{-1}, \quad \Delta S_{\text{fus}} = 1.22 \text{ kJ (kg} \cdot \text{K)}^{-1}.$$

8.2.2.1 Case 1: On Earth (T = 293 K)
The total entropy change (ΔS_{total}) for the melting process is calculated as

$$\Delta S_{\text{total}} = \Delta S_{\text{sys}} - \frac{\Delta H_{\text{fus}}}{T_{\text{surr}}}.$$

Substituting the values,
$$\Delta S_{sys} = \Delta S_{fus} = 1.22 \text{ kJ (kg} \cdot \text{K)}^{-1}, \; T_{surr} = 293 \text{ K}, \; \Delta H_{fus} = 334 \text{ kJ kg}^{-1}$$

$$\Delta S_{total} = 1.22 - \frac{334}{293}.$$

Calculate

$$\Delta S_{total} = 1.22 - 1.14 = 0.08 \text{ kJ (kg} \cdot \text{K)}^{-1}.$$

Since $\Delta S_{total} > 0$, the process is *spontaneous* at this temperature.

8.2.2.2 Case 2: On Mars (T = 223 K)
The total entropy change is calculated similarly,

$$\Delta S_{total} = \Delta S_{sys} - \frac{\Delta H_{fus}}{T_{surr}}.$$

Substituting the values,
$$\Delta S_{sys} = 1.22 \text{ kJ (kg} \cdot \text{K)}^{-1}, \; T_{surr} = 223 \text{ K}, \; \Delta H_{fus} = 334 \text{ kJ kg}^{-1}.$$

$$\Delta S_{total} = 1.22 - \frac{334}{223}.$$

Calculate

$$\Delta S_{total} = 1.22 - 1.50 = -0.28 \text{ kJ (kg} \cdot \text{K)}^{-1}.$$

Since $\Delta S_{total} < 0$, the process is *not spontaneous* at this temperature.

8.3 Reformulation in terms of free energy

To simplify the analysis of spontaneity, we introduce the concept of **Gibbs free energy** (G) as a thermodynamic property defined by

$$G = H - TS,$$

where H is the enthalpy, T is the temperature, and S is the entropy of the system.

For processes at constant temperature ($T = $ constant), the change in Gibbs free energy (ΔG) is given by

$$\Delta G = \Delta H - T\Delta S.$$

We now consider the total entropy change of the universe during the process, which consists of the entropy change of the system and that of the surroundings:

$$\Delta S_{total} = \Delta S_{sys} + \Delta S_{surr}$$

Assuming that the surroundings act as an ideal thermal reservoir at a constant temperature T_{surr}, and that the only interaction with the surroundings is through heat transfer Q, the entropy change of the surroundings is given by:

$$\Delta S_{surr} = -\frac{Q}{T_{surr}}.$$

Substituting this into the total entropy expression, we obtain:

$$\Delta S_{\text{total}} = \Delta S_{\text{surr}} = -\frac{Q}{T_{\text{surr}}}.$$

Applying the First Law of Thermodynamics for a closed system, the heat exchanged can be written as:

$$Q = \Delta H_{\text{sys}} - W$$

which leads to:

$$\Delta S_{\text{total}} = \Delta S_{\text{sys}} - \frac{\Delta H_{\text{sys}} - W}{T_{\text{surr}}}.$$

Multiplying both sides of the equation by T_{surr}, we have:

$$T_{\text{surr}}\Delta S_{\text{total}} = T_{\text{surr}}\Delta S_{\text{sys}} - \left(\Delta H_{\text{sys}} - W\right).$$

Now, under the assumption that the system and surroundings are at the same temperature, i.e., $T_{\text{surr}} = T$, this expression simplifies to:

$$T\Delta S_{\text{total}} = T\Delta S_{\text{sys}} - \left(\Delta H_{\text{sys}} - W\right).$$

Substituting the definition of $\Delta G = \Delta H_{\text{sys}} - T\Delta S_{\text{sys}}$, we obtain

$$T\Delta S_{\text{total}} = -\Delta G + W.$$

From the **second law of thermodynamics**, the total entropy change ΔS_{total} for a process must satisfy

$$\Delta S_{\text{total}} \geqslant 0.$$

This condition implies that

$$T\Delta S_{\text{total}} \geqslant 0.$$

Rearranging the earlier equation

$$-\Delta G + W \geqslant 0 \text{ or equivalently } \Delta G \leqslant W.$$

Thus, the Gibbs free energy change (ΔG) serves as the upper limit of the useful work (W) that can be extracted from a process under constant temperature and pressure conditions. When the process is **reversible**, $\Delta S_{\text{total}} = 0$, and the equality holds:

$$\Delta G = W.$$

For **irreversible processes**, where $\Delta S_{\text{total}} > 0$, the inequality holds strictly:

$$\Delta G < W.$$

Finally, when no useful work is done ($W = 0$) the condition for spontaneity ($\Delta S_{\text{total}} > 0$) becomes

$$\Delta G < 0.$$

This reformulation shows that for a process at constant temperature, the spontaneity of the process is determined by the Gibbs free energy. Specifically:

- If $\Delta G < 0$, the process is **spontaneous**.
- If $\Delta G > 0$, the process is **non-spontaneous**.

8.4 Heat transfer and the role of work in spontaneity

The concepts of spontaneity and entropy can be extended to processes involving heat transfer and work, such as those in heat engines or refrigerators. Let us analyse three distinct cases to understand how spontaneity, entropy changes, and work interplay.

8.4.1 Case 1: Spontaneous heat transfer (without work)

When energy transfers as heat from a warm reservoir to a cold reservoir (figure 8.1), the process occurs spontaneously. This process is governed by the second law of thermodynamics, as it increases the total entropy of the system and surroundings ($\Delta S_{\text{total}}>0$):

- The warm reservoir loses heat (Q_{warm}), reducing its entropy.
- The cold reservoir gains heat (Q_{cold}), increasing its entropy by a greater amount, due to its lower temperature.
- As a result, the net entropy change is positive:

$$\Delta S_{\text{total}} = \Delta S_{\text{warm}} + \Delta S_{\text{cold}} > 0.$$

This **irreversible process** is a natural flow of energy, requiring no work input or output.

8.4.2 Case 2: Spontaneous heat transfer with work generation (heat engine)

If a heat engine operates between the same warm and cold reservoirs (figure 8.1), it harnesses part of the energy transferred as heat to perform work. In this idealized case (no irreversibilities), the total entropy change is zero ($\Delta S_{\text{total}} = 0$):

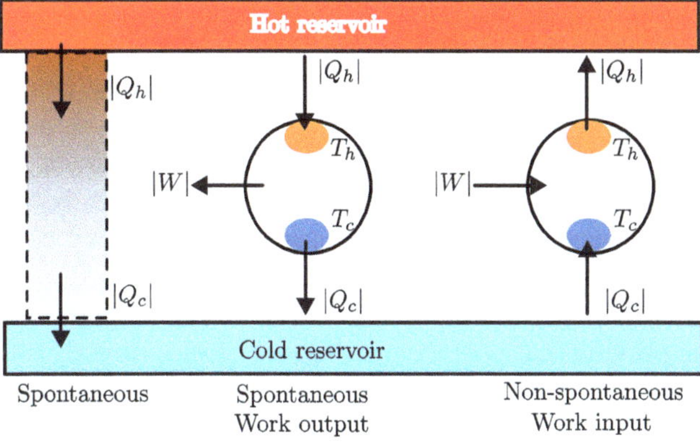

Figure 8.1. Illustration of heat transfer between hot and cold reservoirs. Three scenarios are depicted: (1) spontaneous heat flow from hot to cold (left), (2) spontaneous heat engine operation generating work (center), and (3) non-spontaneous refrigeration requiring work input (right).

- Heat Q_{in} flows from the warm reservoir to the engine.
- Part of this energy is converted into work (W_{out}).
- The remainder is rejected as heat (Q_{out}) to the cold reservoir.
- The entropy gain in the cold reservoir offsets the entropy loss in the warm reservoir, resulting in

$$\Delta S_{total} = \Delta S_{warm} + \Delta S_{cold} = 0.$$

In practice, real engines have irreversibilities (e.g. friction, heat transfer), leading to a slightly positive total entropy change ($\Delta S_{total} > 0$). Even so, the entropy increase is less than in the case of spontaneous heat transfer without work, as the engine utilizes some of the heat flow to produce work.

8.4.3 Case 3: Non-spontaneous heat transfer (work input required)

When energy is transferred from a cold reservoir to a warm reservoir (figure 8.1), the process is **non-spontaneous** because it decreases the total entropy of the system and surroundings ($\Delta S_{total} < 0$) if left to occur naturally. However, this process can be achieved by performing work, as in the case of refrigerators or heat pumps:

- Work (W_{in}) is applied to drive heat (Q_{cold}) from the cold reservoir to the warm reservoir.
- The system's total entropy change remains non-negative because the work done compensates for the entropy reduction in the heat transfer.

For a refrigerator operating ideally:

$$\Delta S_{total} = \Delta S_{cold} + \Delta S_{warm} = 0.$$

In reality, irreversibilities make $\Delta S_{total} > 0$, requiring slightly more work to overcome these inefficiencies.

8.4.4 Summary of the three cases

1. **Spontaneous without work**: Energy transfers naturally from warm to cold, increasing total entropy ($\Delta S_{total} > 0$).
2. **Spontaneous with work generation**: Heat engines harness part of the heat transfer to produce work. Ideally, $\Delta S_{total} = 0$; practically, $\Delta S_{total} > 0$.
3. **Non-spontaneous (work required)**: Energy transfers from cold to warm with work input. Ideally, $\Delta S_{total} = 0$; practically, $\Delta S_{total} > 0$.

Note: Unlike the previous cases of ice melting, we cannot directly use the Gibbs free energy concept (ΔG) in these scenarios because they involve **cyclic processes**. Gibbs free energy applies to changes in a system at constant temperature and pressure, whereas cycles involve multiple stages with varying conditions. Instead, the analysis relies on entropy changes and energy balances over the entire cycle.

8.5 Gas expansion, compression: entropy, and Gibbs free energy

This example combines the concepts of entropy and Gibbs free energy to analyse gas expansion and compression.

8.5.1 Case 1: Spontaneous expansion (no work)

In this scenario, gas expands freely from a high-pressure reservoir to the ambient. This is a **spontaneous process** because it occurs naturally without requiring work input.

- Entropy discussion:
  - As the gas expands, its internal energy decreases, and its entropy increases ($\Delta S_{\text{system}} > 0$) due to the increase in volume and dispersal of energy.
  - There is no heat exchange ($Q = 0$), so the surroundings do not gain or lose entropy ($\Delta S_{\text{surroundings}} = 0$).
  - The total entropy change is positive:

$$\Delta S_{\text{total}} = \Delta S_{\text{system}} + \Delta S_{\text{surroundings}} > 0.$$

- Gibbs free energy:
  - The Gibbs free energy change for the system is **negative**, reflecting the spontaneity of the process:

$$\Delta G < 0.$$

  - Since there is no mechanism to harness the released energy, no useful work is produced ($W = 0$).

8.5.2 Case 2: Controlled expansion with work generation

Here, the gas expands through a turbine, extracting useful work. The process consists of two stages: adiabatic expansion and heat exchange with the surroundings.

- **Stage 1**. Adiabatic expansion in the turbine:
  - The gas expands through the turbine, cooling as it does work on the turbine.
  - Since the process is adiabatic, no heat is exchanged ($Q = 0$), and the entropy of the gas remains constant:

$$\Delta S_{\text{system}} = 0 \text{ (adiabatic and ideal)}.$$

  - The gas's internal energy decreases, and its volume increases.
- **Stage 2**. Heat exchange with the surroundings:
  - After expansion, the gas passes through a heat exchanger where it absorbs heat from the surroundings to restore its temperature to the initial state.
  - During this stage, the entropy of the gas increases ($\Delta S_{\text{system}} > 0$), while the surroundings lose entropy ($\Delta S_{\text{surroundings}} < 0$).

- Total entropy change:
  - For an *ideal process*, the entropy gained by the gas exactly offsets the entropy lost by the surroundings, so the total entropy change is zero:

  $$\Delta S_{\text{total}} = 0.$$

  - *In practice*, irreversibilities such as friction and turbulence result in entropy generation, causing a slight increase in total entropy:

  $$\Delta S_{\text{total}} > 0.$$

- Gibbs free energy:
  - The maximum work obtainable from the system corresponds to the Gibbs free energy change:

  $$W_{\text{max}} = -\Delta G.$$

  - *In practice*, irreversibilities reduce the actual work output:

  $$W_{\text{actual}} < -\Delta G.$$

8.5.3 Case 3: Non-spontaneous compression

In this scenario, work is required to compress the gas from ambient pressure back into the high-pressure reservoir. As case 2, the process consists of two stages: adiabatic compression and heat rejection to the surroundings.

- **Stage 1**. Adiabatic compression in the compressor:
  - The gas is compressed, increasing its internal energy and temperature.
  - Since the process is adiabatic, no heat is exchanged ($Q = 0$), and the entropy of the gas remains constant:

  $$\Delta S_{\text{system}} = 0 \text{ (adiabatic and ideal)}.$$

  - The gas's volume decreases, and its pressure increases.
- **Stage 2**. Heat rejection to the surroundings:
  - After compression, the gas passes through a heat exchanger where it releases heat to the surroundings, returning to its initial temperature.
  - During this stage, the gas loses entropy ($\Delta S_{\text{system}} < 0$), while the surroundings gain entropy ($\Delta S_{\text{surroundings}} > 0$).

- Total entropy change:
  - For an *ideal process*, the entropy lost by the gas is exactly offset by the entropy gained by the surroundings, so the total entropy change is zero:

  $$\Delta S_{\text{total}} = 0.$$

  - *In practice*, irreversibilities such as heat generation and friction lead to a slight increase in total entropy:

  $$\Delta S_{\text{total}} > 0.$$

- Gibbs free energy:
  - The minimum work required to compress the gas corresponds to the Gibbs free energy change:

$$W_{\min} = \Delta G.$$

  - *In practice*, irreversibilities increase the actual work required:

$$W_{\text{actual}} > \Delta G.$$

8.5.4 Why Gibbs free energy is 'free' energy

Gibbs free energy represents the energy available to do *useful work* at constant temperature:
- **Case 2** (expansion): $-\Delta G$ represents the maximum useful work the system can deliver.
- **Case 3** (compression): ΔG represents the minimum work required to drive the process.

It is called 'free' energy because it excludes the heat transfer, focusing solely on the energy that can be utilized or must be supplied.

8.5.5 Summary of the three cases

- **Case 1** (spontaneous expansion):
  $\Delta S_{\text{total}} > 0$, $\Delta G < 0$, and no work is produced ($W = 0$).
- **Case 2** (controlled expansion):
  $\Delta S_{\text{total}} = 0$ ideally (>0 in practice), $\Delta G < 0$, and work is generated ($W_{\text{actual}} < -\Delta G$).
- **Case 3** (non-spontaneous compression):
  $\Delta S_{\text{total}} = 0$ ideally (>0 in practice), $\Delta G > 0$, and work is required ($W_{\text{actual}} > \Delta G$).

8.6 Treating physical and chemical processes systematically

Thermodynamic principles allow us to analyse both **physical** and **chemical** processes systematically, determining their spontaneity, direction, and associated energy exchanges. By expressing such processes as reactions, we can use the tools of thermodynamics to calculate the heat transfer, work involved, and the spontaneity of the process.

8.6.1 Physical processes as reactions

A familiar example of a physical process is the **melting of ice**, which can be expressed as

$$\text{Ice} \rightarrow \text{Water}.$$

This representation allows us to treat melting as a 'reaction', just as we would with a chemical process. By applying thermodynamic principles, we can determine whether the process is spontaneous. For instance:

- At 293 K (20 °C), melting is spontaneous because the Gibbs free energy change is negative ($\Delta G < 0$).
- At -10 °C, freezing is spontaneous as the reverse reaction (Water $\to$ Ice) has $\Delta G < 0$.

Through this approach, we can quantify the heat transfer, entropy changes, and even the maximum or minimum work involved.

8.6.2 Chemical processes as reactions

In addition to physical processes, chemical reactions can also be analysed systematically. For example:

1. **Formation of rust** (engineering example):

$$4Fe + 3O_2 \to 2Fe_2O_3.$$

   This reaction occurs spontaneously in the presence of oxygen and moisture, leading to the formation of rust—a process of interest in material science and corrosion engineering.

2. **Combustion of methane** (energy example):

$$CH_4 + 2O_2 \to CO_2 + 2H_2O.$$

   This reaction is of paramount importance in energy engineering, as it represents the combustion of natural gas to produce energy.

3. **Formation of water** (fuel cell example):

$$2H_2 + O2 \to 2H_2O.$$

   This process is the basis of hydrogen fuel cells, which generate electricity through a controlled reaction between hydrogen and oxygen.

These examples, drawn from physics and engineering contexts, allow us to systematically analyse chemical processes to determine whether they are spontaneous and calculate their energy changes.

8.6.3 Spontaneity and systematic analysis

The central thermodynamic question is: *is the process spontaneous?*

To answer this question, we express the process as a reaction and calculate:

1. **Reaction enthalpy** (ΔH_r):

$$\Delta H_r = \overset{\text{products}}{\underset{i}{\sum}} n_i H_i - \overset{\text{reactants}}{\underset{i}{\sum}} n_i H_i.$$

   This represents the total energy change of the system ($Q + W$) due to the reaction, including contributions from bond breaking and formation.

2. **Reaction entropy** (ΔS_r):

$$\Delta S_r = \sum_i^{\text{products}} n_i S_i - \sum_i^{\text{reactants}} n_i S_i.$$

The entropy change tells us about the energy dispersal during the reaction.

3. **Reaction Gibbs free energy** (ΔG_r):

$$\Delta G_r = \sum_i^{\text{products}} n_i G_i - \sum_i^{\text{reactants}} n_i G_i.$$

The sign of ΔG_r determines whether the process is spontaneous ($\Delta G_r < 0$), non-spontaneous ($\Delta G_r > 0$), or at equilibrium ($\Delta G_r = 0$).

For example:
- A **spontaneous reaction** will proceed in the forward direction (e.g. ice melting at 293 K).
- A **non-spontaneous reaction** requires external energy (e.g. water freezing at 20 °C).
- At **equilibrium**, both forward and reverse reactions occur at the same rate, as seen in reversible processes.

8.6.4 Heat transfer and work in reactions

By systematically representing a process as a reaction, we can determine the associated energy interactions, including heat transfer (Q) and work (W):

1. **Heat transfer** (Q): At constant pressure, the heat transfer for a reaction depends on the enthalpy and work interactions. If no other forms of work (W_{other}) are involved, the heat transfer equals the reaction enthalpy:

$$Q = \Delta H_r \text{ (if } W_{\text{other}} = 0).$$

However, in cases like a **hydrogen fuel cell**, electrical work is extracted as part of the process. In such cases, the relationship changes:

$$Q = T\Delta S_r.$$

Here:
- ΔS_r is the entropy change of the reaction.
- T is the temperature of the system, assumed constant.

This equation highlights that the heat transfer depends on the entropy change of the reaction, rather than the enthalpy directly. For instance, in a fuel cell, the electrical work extracted reduces the energy available as heat, making $Q \neq \Delta H_r$.

2. **Work** (W): The Gibbs free energy change determines the work interactions:
   - **Maximum work (expansion processes)**: The maximum useful work delivered is

$$W_{\text{max}} = -\Delta G_r.$$

This occurs in reversible processes, where all energy from the reaction is efficiently converted to work.

- ○ **Minimum work (compression processes)**: The minimum work required to drive a non-spontaneous reaction is

$$W_{\min} = \Delta G_r.$$

In practical processes, irreversibilities lead to deviations from these ideal values.

8.6.5 Summary

- For reactions without work interactions ($W_{\text{other}} = 0$), $Q = \Delta H_r$.
- For reactions with significant work interactions, $Q = T\Delta S_r$, with the remaining energy extracted as work ($-\Delta G_r$).
- The systematic treatment of ΔH_r, ΔS_r, and ΔG_r provides a complete picture of energy and entropy changes, enabling precise predictions of heat transfer, work, and spontaneity.

8.7 Distinguishing between the Gibbs free energy of a substance and a reaction

In thermodynamics, understanding the behavior of substances and reactions is fundamental when estimating the properties of a system under varying conditions. One crucial concept is Gibbs free energy, which helps determine whether a process will proceed spontaneously. However, it is important to distinguish between the Gibbs free energy of a substance and the Gibbs free energy of a reaction, as their interpretations and implications differ significantly. In this section, we will explore these differences using two key examples: the Gibbs free energy of a single substance (water) in different phases and the Gibbs free energy of a reaction, specifically the phase change of water from liquid to vapor.

8.7.1 Gibbs free energy of a substance

The Gibbs free energy of a substance describes how its energy varies with temperature and pressure under equilibrium conditions. It provides insights into the phase transitions and the stability of the substance at different conditions. The behavior of Gibbs free energy for water across its various phases—ice, liquid water, and steam—can be visualized in figure 8.2.

The graph shows the Gibbs free energy (G) of water as a function of temperature (T) at a constant pressure of 1 bar. In this plot, we can observe three distinct lines representing the different phases of water: ice, liquid water, and steam. The graph clearly shows the following:

1. **The negative slope of each curve**: This reflects the fact that the entropy (S) of the system is positive. As temperature increases, the entropy increases, and since the Gibbs free energy is defined as $G = H - TS$, the negative slope indicates that the system's Gibbs free energy decreases with increasing temperature at a constant pressure.

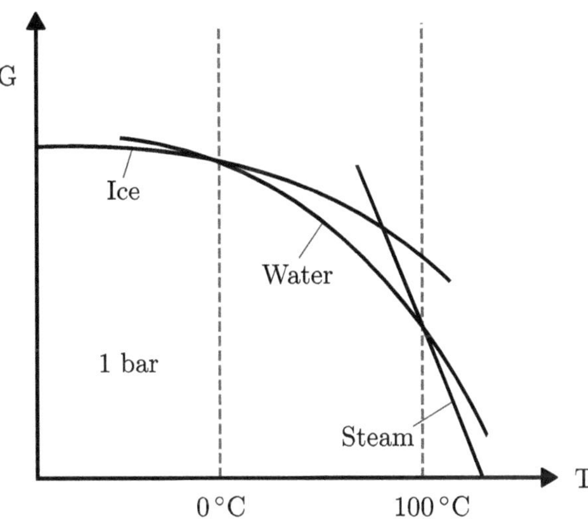

Figure 8.2. The behaviour of Gibbs free energy for water across its various phases.

2. **The curve is not a straight line**: The curves for each phase (ice, water, steam) are not linear because the enthalpy (H) is also a function of temperature. In other words, as the temperature changes, the heat capacity (Cp) of the substance changes, which affects the enthalpy and leads to the non-linear behavior of the Gibbs free energy curve.

The intersection of these curves represents the phase transitions. For instance, at 0 °C, the curve for ice intersects the curve for water, indicating the melting point, and at 100 °C, the curve for liquid water intersects with the steam curve, indicating the boiling point.

8.7.2 Gibbs free energy of a reaction

In contrast, the Gibbs free energy of a reaction represents the overall change in free energy during a chemical or physical transformation, such as a phase transition or a chemical reaction. This is not simply the Gibbs free energy of a substance but rather the difference in Gibbs free energy between the reactants and products, as the reaction progresses.

Let us now consider an example: the phase transition of water from liquid to vapor. The second graph (figure 8.3) shows the Gibbs free energy of the reaction H_2O (l) $\rightarrow$ H_2O (g), which is the vaporization of water at different temperatures and pressures. This process is explained in detail in the next chapter.

This graph is characteristic of a phase change reaction, where the Gibbs free energy is calculated for the system transitioning between two states—liquid and vapor. The important details are:

1. **The decrease in Gibbs free energy as temperature increases**: As we increase the temperature, the Gibbs free energy of the vapor phase (G_{vapor}) becomes lower

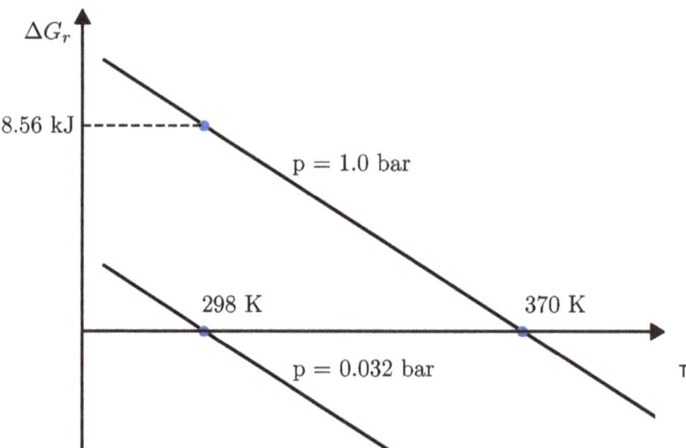

Figure 8.3. The Gibbs free energy of the reaction H_2O (l) $\rightarrow$ H_2O (g).

than that of the liquid phase (G_{liquid}), indicating that the vapor phase becomes more stable at higher temperatures.

2. **The transition point**: At the boiling point, the two phases—liquid and vapor— have the same Gibbs free energy, meaning the system is at equilibrium. At this point, any additional heat added to the system will cause the phase transition to occur without a change in temperature, leading to the vaporization of the liquid water.

3. **The pressure dependence**: The phase change from liquid to vapor is also influenced by pressure. In the graph, the transition occurs at different temperatures for different pressures, demonstrating how the Gibbs free energy of the reaction depends on both temperature and pressure. For example, at higher pressures, the boiling point of water increases, which is consistent with the Clausius–Clapeyron relation.

8.7.3 Key differences between Gibbs free energy of a substance and a reaction

- **Substance**: The Gibbs free energy of a substance, represents its potential energy and its tendency to change phase based on temperature. The curve (figure 8.2) reflects how the substance behaves at a given pressure and temperature, and the intersection points of the curves signify phase transitions. The negative slope indicates that, as temperature increases, the system's entropy increases, and the Gibbs free energy decreases.

- **Reaction**: On the other hand, the Gibbs free energy of a reaction, reflects the change in free energy as the system progresses from one phase (liquid) to another (vapor). This curve (figure 8.3) shows the relationship between the Gibbs free energy of the reactants and products during the transformation, and it demonstrates how external factors such as pressure and temperature affect the spontaneity of the phase change or reaction.

In the case of the phase transition of water from liquid to vapor, the reaction is spontaneous at higher temperatures because the Gibbs free energy of the vapor phase becomes lower than that of the liquid phase, and the transition occurs when their Gibbs free energies become equal.

The key distinction between the Gibbs free energy of a substance and that of a reaction lies in what they represent. The Gibbs free energy of a substance reflects how the energy of a single phase evolves with temperature and pressure, and it allows us to predict phase transitions. In contrast, the Gibbs free energy of a reaction reflects the change in free energy during a transformation between different phases or chemical states, indicating whether the transformation is spontaneous or not under given conditions.

8.8 The Clausius–Clapeyron relation and phase transitions

In thermodynamics, the Clausius–Clapeyron relation is crucial for understanding phase transitions, particularly in the context of vaporization, sublimation, and melting. It links the change in Gibbs free energy (ΔG) with changes in pressure and temperature during a phase transition. In this section, we will focus on the derivation and application of the Clausius–Clapeyron relation for phase transitions.

8.8.1 Derivation of the Clausius–Clapeyron relation

Consider a phase transition between two phases of a substance, for example, liquid to vapor, at constant temperature and pressure. At equilibrium, the change in Gibbs free energy (ΔG) for this transition must be zero:

$$\Delta G_r = \Delta H_r - T\Delta S_r = 0,$$

where:
- ΔH_r is the enthalpy change of the phase transition (latent heat),
- ΔS_r is the entropy change of the transition, and
- T is the temperature at which the phase transition occurs.

This simplifies to

$$\Delta H_r = T\Delta S_r.$$

For a phase transition occurring at constant temperature, the change in Gibbs free energy between the two phases can be expressed in terms of the volume change (ΔV_r) and pressure (p):

$$d\Delta G_r = -\Delta S_r dT + \Delta V_r dp = 0.$$

Since we are considering the equilibrium condition for the phase transition, the total differential change in Gibbs free energy is zero. This implies that the change in Gibbs free energy is related to both the entropy and volume changes during the phase transition.

Now, rearranging the equation, we obtain

$$\frac{\Delta S_r}{\Delta V_r} = \frac{dp}{dT}.$$

To isolate the relationship between pressure and temperature, we rewrite this equation in a form that relates the vapor pressure (p) to the temperature (T). Using the fact that $\Delta H_r = T\Delta S_r$, we can express the latent heat of the phase transition as

$$\Delta H_r = T\Delta V_r \frac{dp}{dT}.$$

Now, to proceed further, let us treat the vapor as an ideal gas. The ideal gas law is given by

$$pV = nRT.$$

For the vapor phase, we can approximate the volume change ΔV_r as the volume of the vapor phase, since the volume of the liquid phase is negligible compared to that of the vapor phase. Thus, $\Delta V_r \approx \frac{RT}{p}$, which is derived from the ideal gas law.

Substituting this into the earlier equation,

$$\Delta H_r = T\left(\frac{RT}{p}\right)\frac{dp}{dT}.$$

Simplifying,

$$\Delta H_r = \frac{RT^2}{p}\frac{dp}{dT}.$$

Now, let us isolate $\frac{dp}{dT}$:

$$\frac{dp}{dT} = \frac{p\Delta H_r}{RT^2}.$$

Taking the natural logarithm of both sides to express the pressure–temperature relationship, we obtain

$$\frac{d \ln p}{dT} = \frac{\Delta H_r}{RT^2}.$$

This is the Clausius–Clapeyron relation, which describes how the vapor pressure p of a substance changes with temperature T.

8.8.2 Integration of the Clausius–Clapeyron relation

To integrate this equation, we assume that the latent heat of the phase transition (ΔH_r) is constant over the temperature range, which is often a valid approximation for small temperature intervals. The integrated form of the Clausius–Clapeyron relation becomes

$$\ln\left(\frac{p_2}{p_1}\right) = \frac{\Delta H_r}{R}\left(\frac{1}{T_1} - \frac{1}{T_2}\right),$$

where:
- p_1 and p_2 are the vapor pressures at temperatures T_1 and T_2, respectively.

This form of the Clausius–Clapeyron relation allows you to calculate the change in vapor pressure (p) for a substance over a given temperature range, given the latent heat of the phase transition. This equation is particularly useful for determining the boiling point of liquids under different pressures, such as predicting the boiling point of water at varying altitudes.

8.9 Thermodynamic analysis of hydrogen reactions

To further illustrate the interplay between enthalpy, entropy, and Gibbs free energy in thermodynamics, we consider three reactions involving hydrogen: combustion of hydrogen, operation of a hydrogen fuel cell, and electrolysis of water. These cases demonstrate spontaneous and non-spontaneous processes and how energy transfers can be quantified.

We focus on **one mole of water** in all reactions to ensure consistency and simplify the analysis (see table 8.1 for a summary of the thermodynamic values).

8.9.1 Case 1: Combustion of hydrogen

- *Reaction*:

$$H_2 \text{ (g)} + 1/2O_2 \text{ (g)} \rightarrow H_2O \text{ (l)}.$$

- *Thermodynamic parameters*:
    - Enthalpy change (ΔH_r),

$$\Delta H_r = -285.8 \text{ kJ mol}^{-1} \text{ of water.}$$

Table 8.1. Summary of the thermodynamic values for hydrogen reactions (per mole of water).

Process	ΔH_r (kJ mol^{-1})	ΔS_r (kJ mol$^{-1} \cdot$ K^{-1})	ΔG_r (kJ mol)$^{-1}$	Spontaneity	Work interaction
Combustion of hydrogen	-285.8	-0.163	-237.2	Spontaneous	No work produced
Hydrogen fuel cell	-285.8	-0.163	-237.2	Spontaneous	$W_{\text{electrical}} = 237.2$ kJ mol^{-1}
Electrolysis of water	$+285.8$	$+0.163$	$+237.2$	Non-spontaneous	$W_{\text{electrical}} = 237.2$ kJ mol^{-1}

- Entropy change (ΔS_r),

$$\Delta S_r = -163 \text{ J mol}^{-1}\,\text{K}^{-1} = -0.163 \text{ kJ mol}^{-1}\,\text{K}^{-1}.$$

- Gibbs free energy change (ΔG_r), at 298 K:

$$\Delta G_r = \Delta H_r - T\Delta S_r.$$

$$\Delta G_r = -285.8 - (298 \times -0.163).$$

$$\Delta G_r = -285.8 + 48.6 = -237.2 \text{ kJ mol}^{-1}.$$

- *Interpretation*:
  - The negative ΔG_r indicates that combustion is *spontaneous*.
  - The reaction is highly exothermic, releasing significant heat.

8.9.2 Case 2: Hydrogen fuel cell

- *Reaction*:

$$H_2 \text{ (g)} + \frac{1}{2}O_2 \text{ (g)} \rightarrow H_2O \text{ (l)}.$$

This reaction occurs electrochemically in a fuel cell to generate electricity.
- *Thermodynamic parameters*:
  - Enthalpy change (ΔH_r),

$$\Delta H_r = -285.8 \text{ kJ mol}^{-1} \text{ of water.}$$

- Entropy change (ΔS_r),

$$\Delta S_r = -0.163 \text{ kJ mol}^{-1}\,\text{K}^{-1}$$

- Gibbs free energy change (ΔG_r),

$$\Delta G_r = -237.2 \text{ kJ mol}^{-1}.$$

- *Energy distribution*:
  - Electrical work: The maximum electrical work equals the Gibbs free energy change,

$$W_{\text{electrical}} = -\Delta G_r = 237.2 \text{ kJ mol}^{-1}.$$

- Heat released: The remaining energy is dissipated as heat,

$$Q = \Delta H_r - W_{\text{electrical}} = -285.8 - (-237.2) = -48.6 \text{ kJ mol}^{-1}.$$

- *Interpretation*:
  - The reaction is *spontaneous*, as $\Delta G_r < 0$.

- The fuel cell efficiently converts part of the energy into electrical work, with the rest released as heat.

8.9.3 Case 3: Electrolysis of water

- *Reaction*:

$$H_2O \ (l) \rightarrow H_2 \ (g) + \frac{1}{2}O_2 \ (g).$$

This is the reverse of the combustion reaction and requires energy input.
- *Thermodynamic parameters*:
  - Enthalpy change (ΔH_r),

$$\Delta H_r = +285.8 \ \text{kJ mol}^{-1} \text{ of water.}$$

  - Entropy change (ΔS_r),

$$\Delta S_r = +0.163 \ \text{kJ mol} \cdot \text{K}^{-1}.$$

  - Gibbs free energy change (ΔG_r),

$$\Delta G_r = \Delta H_r - T\Delta S_r.$$

$$\Delta G_r = 285.8 - (298 \times 0.163).$$

$$\Delta G_r = 285.8 - 48.6 = 237.2 \ \text{kJ mol}^{-1}.$$

- *Energy requirements*:
  - Electrical work: The minimum electrical work required equals the Gibbs free energy change,

$$W_{\text{electrical}} = \Delta G_r = 237.2 \ \text{kJ mol}^{-1}.$$

  - Heat absorbed: Additional heat is absorbed to supply the entropy-related energy,

$$Q = \Delta H_r - W_{\text{electrical}} = 285.8 - 237.2 = 48.6 \ \text{kJ mol}^{-1}.$$

- *Interpretation*:
  - The positive ΔG_r confirms that electrolysis is *non-spontaneous* and requires external energy.
  - Efficient electrolysis minimizes electrical energy input by utilizing heat from the surroundings.

8.10 Phase change process: graphite to diamond and vice versa

The transformation between graphite and diamond is a fascinating phase change process that exemplifies the interplay of thermodynamic properties such as enthalpy (ΔH), entropy (ΔS), and Gibbs free energy (ΔG). Both graphite and diamond are

Table 8.2. Summary of the thermodynamic values for the graphite to diamond and diamond to graphite reactions.

Reaction	ΔH (kJ mol^{-1})	ΔS (kJ mol^{-1} ° K^{-1})	ΔG (kJ mol^{-1})	Spontaneity
Graphite to diamond	+1.9	−0.0034	+ 2.91	Non-spontaneous
Diamond to graphite	−1.9	+0.0034	−2.91	Spontaneous

solid phases of carbon, but their thermodynamic stability depends on the pressure and temperature conditions (see table 8.2 for a summary of the thermodynamic values).

8.10.1 Graphite to diamond ($C_{graphite} \rightarrow C_{diamond}$)

- *Reaction*:

$$C_{graphite} \rightarrow C_{diamond}.$$

- *Thermodynamic parameters, at standard conditions (298 K, 1 atm)*:
  - Enthalpy change (ΔH),

$$\Delta H = +1.9 \text{ kJ mol}^{-1}.$$

    Diamond has slightly higher enthalpy than graphite, indicating it is less stable energetically under standard conditions.
  - Entropy change (ΔS),

$$\Delta S = -3.4 \text{ J mol K}^{-1} = -0.0034 \text{ kJ mol}^{-1} \text{ K}^{-1}.$$

    The negative entropy change reflects the increased structural order of diamond compared to graphite.
  - Gibbs free energy change (ΔG),

$$\Delta G = \Delta H - T\Delta S.$$

    At 298 K

$$\Delta G = 1.9 - (298 \times -0.0034) = 1.9 + 1.01 = +2.91 \text{ kJ mol}^{-1}.$$

    The positive ΔG indicates that the transformation of graphite to diamond is *non-spontaneous* under standard conditions.
- *Interpretation*:
  - Graphite is the more stable phase at standard conditions (1 atm, 298 K).

8.10.2 Diamond to graphite ($C_{diamond} \rightarrow C_{graphite}$)

- *Reaction*:

$$C_{diamond} \rightarrow C_{graphite}.$$

- *Thermodynamic parameters*:
  - Enthalpy change (ΔH),

$$\Delta H = -1.9 \, \text{kJ mol}^{-1}.$$

  The reverse transformation releases energy as graphite is more stable.
  - Entropy change (ΔS),

$$\Delta S = +0.0034 \, \text{kJ mol}^{-1} \, \text{K}^{-1}.$$

  The positive entropy change reflects the increase in disorder when diamond converts to graphite.
  - Gibbs free energy change (ΔG),

$$\Delta G = -2.91 \, \text{kJ mol}^{-1}.$$

- *Interpretation*:
  - The negative ΔG indicates that diamond spontaneously converts to graphite at standard conditions.
  - However, the process is extremely slow because of the high activation energy required to break the strong covalent bonds in diamond's crystalline structure. This is why diamonds remain stable over geological timescales.

8.11 Helmholtz free energy

The Helmholtz free energy (F) is another thermodynamic potential, defined as

$$F = U - TS,$$

where U is the internal energy, T is the temperature, and S is the entropy of the system. The Helmholtz free energy is particularly useful for analysing systems under varying volume and temperature conditions.

8.11.1 When to use Helmholtz free energy

Unlike the Gibbs free energy, which is used for processes at varying pressure and temperature, the Helmholtz free energy is applicable when:

1. The system's *volume is varying*.
2. The system's *temperature is varying*.

Under these constraints:

- The change in Helmholtz free energy (ΔF) represents the **maximum useful work** that can be extracted from the system.
- Spontaneity is determined by the sign of ΔF:
  - $\Delta F < 0$: The process is spontaneous.
  - $\Delta F = 0$: The system is at equilibrium.
  - $\Delta F > 0$: The process is non-spontaneous.

8.11.2 Relation to Gibbs free energy

The analysis of processes using Helmholtz free energy follows a similar framework to Gibbs free energy, with the key difference being the constraints:
- Gibbs free energy is suitable for systems at *varying pressure and temperature.*
- Helmholtz free energy is suitable for systems at *varying volume and temperature.*

For example:
- In studying chemical reactions or phase changes at constant pressure (common in open systems such as Earth's atmosphere), Gibbs free energy is the appropriate tool.
- For systems such as confined gases in a sealed, rigid container (constant volume), phase transitions in a closed space, or molecular simulations of small systems, Helmholtz free energy becomes more relevant because it quantifies the maximum work obtainable at constant volume and temperature. Examples include studying gas behavior in sealed chambers, analysing protein folding in solvents, modeling phase transitions in crystalline solids, or exploring the thermodynamics of plasmas and polymers under fixed volume constraints.

8.12 Summary

In this chapter, we developed a systematic understanding of the spontaneity of processes using thermodynamic potentials, particularly Gibbs free energy. Through examples such as ice melting on Earth and Mars, we demonstrated how **Gibbs free energy** changes with temperature and pressure and how it governs the spontaneity of processes. The chapter emphasizes the importance of phase transitions and the relationship between enthalpy, entropy, and free energy in determining whether a process will occur spontaneously.

We also introduced the **Clausius–Clapeyron relation**, which is used to model phase transitions such as liquid–vapor or sublimation transitions. The Clausius–Clapeyron relation provides a direct link between the change in Gibbs free energy and pressure/ temperature changes during phase transitions. This relation was derived and applied to practical examples, such as water boiling at varying altitudes and pressures.

Furthermore, the chapter discussed the **Helmholtz free energy**, which applies to systems at constant volume and temperature, contrasting it with the Gibbs free energy, which is used for constant pressure systems. Helmholtz free energy helps in analysing systems such as confined gases or small systems in simulations.

Finally, through comprehensive examples—such as hydrogen combustion, hydrogen fuel cells, and the electrolysis of water—we illustrated the relationship between Gibbs free energy, enthalpy, entropy, and the maximum work achievable or required in various chemical processes. These examples serve to deepen the understanding of how thermodynamics applies to real-world applications.

By synthesizing the concepts of entropy, enthalpy, Gibbs free energy, and Helmholtz free energy, this chapter provides the theoretical foundation and practical tools necessary for analysing and predicting the behavior of systems in diverse thermodynamic processes.

IOP Publishing

A Classical Thermodynamics Toolkit

Srinivas Vanapalli

Chapter 9

Thermodynamic potentials—applications under non-standard temperature and pressure conditions

9.1 Introduction

In this chapter, we delve deeper into the concept of thermodynamic potentials, focusing on their behavior under varying conditions. The chapter builds upon the foundational thermodynamic principles established earlier, particularly concerning free energy and the spontaneity of processes. It introduces key thermodynamic potentials—internal energy (U), enthalpy (H), and Gibbs free energy (G)—and explores their intricate relationships with temperature, pressure, and volume.

These potentials provide a comprehensive framework for understanding the energy exchanges that occur in both reversible and irreversible processes. Through their partial derivatives with respect to state variables, such as entropy, volume, pressure, and temperature, we can derive important thermodynamic quantities. This enables a systematic approach to analysing various physical processes and reactions, including those that involve phase transitions, chemical reactions, and work performed during expansion or compression.

A central theme in this chapter is the application of these potentials in practical scenarios, especially in processes involving work. We explore the various types of work expressions such as $-pdV$ work and Vdp work, highlighting their differences in thermodynamic contexts. Furthermore, we discuss how to calculate and interpret the Gibbs free energy change for both substances and reactions, extending these concepts to conditions where temperature and pressure deviate from standard values.

doi:10.1088/978-0-7503-6029-6ch9   9-1

9.2 Thermodynamic potentials and their derivatives

Thermodynamic potentials, such as internal energy (U), enthalpy (H), and Gibbs free energy (G), are fundamental properties that describe the energy state of a system. These potentials depend on specific state variables and provide valuable insights into the behavior of systems under different thermodynamic conditions. By exploring their mathematical relationships, we can derive useful expressions that connect physical properties like temperature, pressure, and entropy.

9.2.1 The internal energy $U(S, V)$

The internal energy U is a function of entropy S and volume V. From the thermodynamic identity,

$$dU = TdS - pdV.$$

We see that the partial derivatives of U are

$$\left(\frac{\partial U}{\partial S}\right)_V = T \ \text{ and } \ \left(\frac{\partial U}{\partial V}\right)_S = -p.$$

These relationships indicate that:
- T: The temperature of the system is the partial derivative of internal energy with respect to entropy at constant volume.
- $-p$: The negative pressure is the partial derivative of internal energy with respect to volume at constant entropy.

9.2.2 The enthalpy $H(S, p)$

The enthalpy H is defined as

$$H = U + pV.$$

Differentiating, we obtain

$$dH = dU + d(pV) = TdS - pdV + pdV + Vdp = TdS + Vdp.$$

Here, H is a function of entropy S and pressure p. From the expression

$$dH = TdS + Vdp.$$

The partial derivatives are

$$\left(\frac{\partial H}{\partial S}\right)_p = T \ \text{ and } \ \left(\frac{\partial H}{\partial p}\right)_S = V.$$

These relationships reveal that:
- T: The temperature is the partial derivative of enthalpy with respect to entropy at constant pressure.
- V: The volume of the system is the partial derivative of enthalpy with respect to pressure at constant entropy.

9.2.3 The Gibbs free energy $G(p, T)$

The Gibbs free energy G is defined as

$$G = H - TS.$$

Differentiating,

$$dG = dH - d(TS) = TdS + Vdp - TdS - SdT = Vdp - SdT.$$

Here, G is a function of pressure p and temperature T. From the expression,

$$dG = Vdp - SdT.$$

The partial derivatives are

$$\left(\frac{\partial G}{\partial p}\right)_T = V \text{ and } \left(\frac{\partial G}{\partial T}\right)_p = -S.$$

These relationships indicate that:
- V: The volume is the partial derivative of Gibbs free energy with respect to pressure at constant temperature.
- $-S$: The negative entropy is the partial derivative of Gibbs free energy with respect to temperature at constant pressure.

9.2.4 Summary table

To summarize these relationships, we present table 9.1.

These relationships demonstrate how thermodynamic potentials and their derivatives provide a complete description of the thermodynamic behavior of systems under various constraints. The clarity of this framework simplifies the analysis of practical thermodynamic processes.

Table 9.1. Summary of thermodynamic potentials, their natural independent variables, and the resulting partial derivatives. These identities are fundamental to evaluating temperature, pressure, volume, and entropy from the internal energy U, enthalpy H, and Gibbs free energy G, depending on the chosen representation of the system.

Potential X	Independent variables	$\left(\frac{\partial X}{\partial a}\right)_b$	Result
$U(S, V)$	S, V	$\left(\frac{\partial U}{\partial S}\right)_V$	T
		$\left(\frac{\partial U}{\partial V}\right)_S$	$-p$
$H(S, p)$	S, p	$\left(\frac{\partial H}{\partial S}\right)_p$	T
		$\left(\frac{\partial H}{\partial p}\right)_S$	V
$G(p, T)$	p, T	$\left(\frac{\partial G}{\partial p}\right)_T$	V
		$\left(\frac{\partial G}{\partial T}\right)_p$	$-S$

9.3 Examples to illustrate the use of thermodynamic relationships

This section presents detailed examples based on the thermodynamic relationships $dU = TdS - pdV$, $dH = TdS + Vdp$, and $dG = Vdp - SdT$. These examples demonstrate how thermodynamic potentials are used under specific conditions to calculate energy changes and work.

9.3.1 Example 1: $S = $ constant, $dU = -pdV$

This example highlights an isentropic process ($S = $ constant) where the internal energy change (dU) equals the pV-work done by or on the system.

Scenario: A gas compressor compresses a gas in a piston–cylinder system.

- *Initial and final conditions*: The gas starts at a low pressure and high volume. As the piston compresses the gas, its volume decreases while its pressure increases.
- *Key process assumption*: The process is isentropic ($S = $ constant), meaning no entropy change occurs.
- *Analysis*: $dU = TdS - pdV$. With $S = $ constant, $dS = 0$, simplifying to $dU = -p\, dV$. The internal energy change equals the work done on the gas during compression.

9.3.2 Example 2: $S = $ constant, $dH = Vdp$

This example shows how the change in enthalpy (dH) corresponds to the work associated with pressure changes during an isentropic flow process.

Scenario: A gas compressor increases the pressure of a gas as it flows through the machine.

- *Initial and final conditions*: The gas enters the compressor at a low pressure and exits at a higher pressure. The flow process is steady and involves zero heat exchange.
- *Key process assumption*: The process is isentropic ($S = $ constant).
- *Analysis*: $dH = TdS + Vdp$. With $S = $ constant, $dS = 0$, simplifying to: $dH = Vdp$. The increase in enthalpy corresponds to the work done by the compressor to increase the gas pressure.

9.3.3 Example 3: $T = $ constant, $dG = Vdp$

This example illustrates that under constant temperature conditions, the change in Gibbs free energy (dG) equals the flow work resulting from pressure changes.

Scenario: A gas compressor increases the pressure of a gas as it flows through the machine.

- *Initial and final conditions*: The gas enters the compressor at a low pressure and exits at a higher pressure. The flow process is steady and involves zero temperature change.
- *Key process assumption*: The temperature at the entry and exit is the same ($T_{\text{entry}} = T_{\text{exit}}$).

- *Analysis*: $dG = Vdp - SdT$. Since the temperature is the same at the entry and exit ($dT = 0$), the equation simplifies to: $dG = Vdp$. The change in Gibbs free energy reflects the work associated with the pressure increase under constant temperature conditions.

9.3.4 Example 4: $p = $ constant, $dH = TdS$

This example demonstrates that for a constant pressure process, the change in enthalpy (dH) corresponds to the heat transfer if no other work is involved.

Scenario: Heating a gas at constant pressure.

- *Initial and final conditions*: The gas is heated, increasing its temperature and entropy while maintaining constant pressure.
- *Key process assumption*: The pressure is constant ($p = $ constant).
- *Analysis*: $dH = TdS + Vdp$. With $p = $ constant, $dp = 0$, simplifying to $dH = TdS$. The enthalpy change corresponds to the heat transfer that increases the system's entropy.

9.4 Understanding the difference between $-pdV$ and Vdp work

In thermodynamics, the work done by or on a system is typically a result of pressure and volume changes. The two main expressions for work, $-pdV$ and Vdp, describe different scenarios. To illustrate these differences clearly, let us consider the sequence of events in a piston–cylinder system where both the volume and pressure of the gas change during the process.

In figure 9.1, the gas is expanding in a piston–cylinder system, pushing the piston outward. This results in an increase in the volume of the gas, and the work done by the system is captured by the expression $-pdV$, where p is the pressure and dV is the change in volume. This represents **positive work** done by the gas on the surroundings. The pressure remains relatively constant during the expansion process, and the system's volume increases as the gas pushes the piston outward. The work is directly proportional to the pressure and the change in volume.

Figure 9.1. A schematic of a piston–cylinder system under the influence of an external weight m. The weight exerts a constant force on the piston, maintaining a fixed pressure inside the system. This setup is often used to conceptualize reversible processes and study thermodynamic work interactions at constant pressure.

9.4.1 Understanding the Vdp work using figure 9.2

9.4.1.1 Figure 9.2, panel 1: Initial set-up (valve closed)

In this initial scenario, the system consists of a piston with the valve closed. The pressure in the system is at P_{in}, and gas is not allowed to leave because the valve is closed. The system starts at an initial state with a certain volume.

- *Description*: The pressure is constant at P_{in}, and no work is done at this stage, as the volume is not changing.
- *Work expression*: No work is done in this configuration because there is no volume change, and the system is static.

9.4.1.2 Figure 9.2, panel 2: Gas expansion (piston moving right, valve still closed)

In this panel, the piston moves outward, expanding the gas inside. The pressure remains constant at P_{in} as the gas pushes against the piston. The valve is still closed, so no gas can exit the system.

- *Description*: As the piston moves outward, the system's volume increases, and the pressure remains constant at P_{in}.
- *Work expression*: The work done by the gas is given by the expression $W = -\int_p dV$. The work is *negative* since the volume increases ($dV > 0$) during expansion.

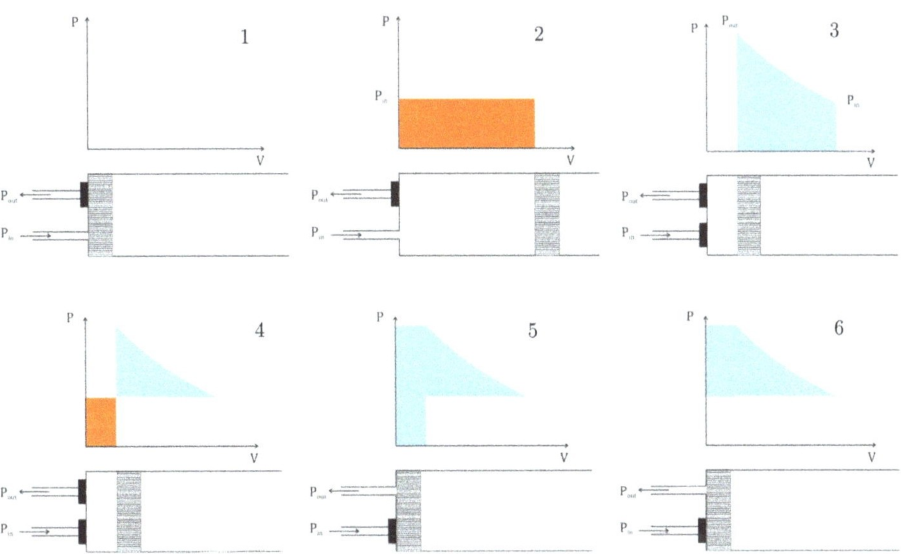

Figure 9.2. Six different quasistatic processes involving an ideal gas in a piston–cylinder system with open valves. Each configuration results in boundary work. The difference lies in how the piston moves and how the inlet and outlet valves are operated. In all cases, the pressure inside the cylinder remains nearly uniform, and the expansion or compression is slow enough to ensure a quasistatic process.

9.4.1.3 *Figure 9.2, panel 3: Compression (piston moving left, valve closed)*

In this step, the piston starts moving inward, compressing the gas. The volume decreases, and the pressure increases within the system since the valve is still closed, and the gas cannot escape.

- *Description*: The gas is compressed, and the volume decreases while the pressure increases. The valve is closed, so the gas remains trapped inside the system.
- *Work expression*: The work done is still represented by $W = -\int_p dV$, but since the volume decreases ($dV < 0$), the work done on the system is *positive*.

9.4.1.4 *Figure 9.2, panel 4: Valve opens, no change in work*

In this panel, we are showing the work done so far, but nothing new happens at this point. The system is still in the same state, with the gas having been compressed up to this point. The work done thus far is represented by the area under the curve.

- *Description*: There is no further change in pressure or volume in this step; it is just showing the work that has accumulated up to this point.
- *Key insight*: We are simply displaying the accumulated work until the valve opens. The graph shows the work so far, but no new energy transfer occurs.

9.4.1.5 *Figure 9.2, panel 5: Valve opens, gas moves out at constant pressure*

In this step, the valve opens, and the gas begins to move out of the system. The pressure remains constant during this process, and the gas leaves the system as the piston continues to move inward.

- *Description*: The valve opens, and the gas starts exiting the system. The pressure remains constant as the gas leaves, and the system now undergoes a flow process. The volume decreases as the gas moves out, and the work done is related to the pressure–volume relationship, represented by the integral $\int_V dp$.
- *Key insight*: In this case, the gas is expelled at a constant pressure, and the work done during this phase is captured by the flow work, where Vdp characterizes the pressure change while the volume remains approximately constant.

9.4.1.6 *Figure 9.2, panel 6: Net work as Vdp area under the graph*

In this final step, panel 6 shows the net work done during the process. The figure represents the area under the pressure–volume curve, capturing the net flow work done as the gas exits the system.

- *Description*: No new changes occur in this step; it simply shows the total work accumulated up to this point. The net work is represented by the area under the pressure–volume curve, reflecting the integral $\int_V dp$ from the pressure drop while the gas exits the system.
- *Key insight*: This figure highlights the net work done during the entire process, as reflected in the area under the curve, but no new events or changes occur in this step.

9.4.1.7 *Summary*

This sequence demonstrates the difference between $-pdV$ and Vdp work:

1. $-pdV$ describes work done during compression or expansion in a closed system where both pressure and volume change.

2. Vdp describes work done during a flow process where the pressure changes, but the volume remains nearly constant as gas exits or enters the system.

The panels in figure 9.2 provide a clear illustration of how the work is done in different thermodynamic scenarios.

9.5 Distinguishing between substance and reaction: thermodynamic properties across conditions

In thermodynamics, understanding the behavior of substances and reactions is fundamental, especially when estimating the properties of a system at different temperatures and pressures. A **single substance** refers to the study of the properties of a material, such as water, at various conditions. These properties include Gibbs free energy (G), enthalpy (H), and entropy (S), and they provide insights into how the material behaves as temperature and pressure change. For example, we can observe how the enthalpy of water changes when its temperature is increased or when it is subjected to different pressures. In this context, we are not concerned with any transformation or phase change of the material, just how its intrinsic properties evolve with the environment.

On the other hand, a **reaction** refers to a process where the system undergoes a transformation from one phase or state to another. A common example is the phase change of water from liquid to gas during boiling. During such a reaction, the system's enthalpy, entropy, and Gibbs frce energy change as the substance undergoes a phase transition due to changes in temperature and pressure. This process involves energy exchange with the surroundings and may also be influenced by external conditions, such as heating or cooling rates.

Understanding the difference between a substance and a reaction is crucial because while substances are studied individually for how their properties respond to changing conditions, reactions focus on how entire systems evolve during transitions between different states. This subtle difference is important and often not well clarified in standard texts. In our book, we aim to provide a clear explanation of both cases and introduce tools to estimate the thermodynamic properties—enthalpy, entropy, and Gibbs free energy—for both individual substances and chemical or physical reactions.

To achieve this, we will present methods for calculating these properties in different environments. For substances, we will show how to determine enthalpy, entropy, and Gibbs free energy at various temperatures and pressures based on known data. For reactions, we will introduce the concept of reaction enthalpy, entropy, and Gibbs free energy, and how they vary with temperature and pressure during phase transitions or chemical processes.

9.6 Calculating enthalpy, entropy, and Gibbs free energy at non-standard pressure and temperature

Up to this point, we have primarily considered reactions under standard conditions of pressure and temperature. However, the thermodynamic principles we have discussed are equally valid for systems under non-standard conditions. This is particularly important when designing systems for environments such as Mars, deep under the Earth's surface, or in a cruising airplane, where the temperature and pressure conditions differ from standard atmospheric values. Other examples include high-pressure reactors, deep-sea environments, or industrial applications operating under varying temperatures and pressures.

Thermodynamic properties such as enthalpy (H), entropy (S), and Gibbs free energy (G) at specific pressures and temperatures can often be extracted from databases. However, for quick estimates, it is helpful to have a method for calculating approximate values of these properties at different pressures and temperatures, based on data from standard conditions. In this section, we will present a procedure for calculating enthalpy, entropy, and Gibbs free energy at any given temperature, both for pure substances and reactions.

9.6.1 Determining approximate enthalpy at any temperature

The heat capacity at constant pressure (C_p) is defined as the rate of change of enthalpy with respect to temperature:

$$C_p = \left(\frac{\partial H}{\partial T}\right)_p.$$

To calculate the enthalpy at a temperature T, we can integrate this expression from a reference temperature T_0:

$$H(T) = H(T_0) + \int_{T_0}^{T} C_p(T)\, dT.$$

If we assume that the heat capacity is constant and independent of temperature, the equation simplifies to

$$H(T) = H(T_0) + C_p(T - T_0).$$

Here, $H(T_0)$ represents the standard enthalpy value at the reference temperature.

9.6.2 Determining approximate entropy at any temperature

Entropy can be defined as

$$dS = \frac{Q}{T} = \frac{C_p(T)\, dT}{T}.$$

Integrating this expression gives the entropy at temperature T:

$$S(T) = S(T_0) + \int_{T_0}^{T} \frac{C_p(T)}{T}\, dT.$$

If we assume that the heat capacity is constant and independent of temperature, this simplifies to

$$S(T) = S(T_0) + C_p \ln\left(\frac{T}{T_0}\right).$$

Here, $S(T_0)$ represents the standard entropy value at the reference temperature.

9.6.3 Determining approximate Gibbs free energy at any temperature

The Gibbs free energy at any temperature is related to the enthalpy and entropy by

$$G(T) = H(T) - T \cdot S(T).$$

Thus, the Gibbs free energy can be calculated by substituting the values of $H(T)$ and $S(T)$ derived above. These equations provide a straightforward way to estimate the Gibbs free energy at different temperatures based on known values at standard conditions.

9.6.4 Calculating the Gibbs free energy of a reaction at any temperature

When considering reactions, we are particularly interested in how the enthalpy, entropy, and Gibbs free energy change with temperature and pressure. For example, consider a general reaction

$$2A + B \longrightarrow 3C + D.$$

The enthalpy of the reaction at standard conditions is given by

$$\Delta H_r(p_0, T_0) = 3H_C(p_0, T_0) + H_D(p_0, T_0) - 2H_A(p_0, T_0) - H_B(p_0, T_0).$$

At any temperature T, the enthalpy change is

$$\Delta H_r(p_0, T) = 3H_C(p_0, T) + H_D(p_0, T) - 2H_A(p_0, T) - H_B(p_0, T).$$

By expanding the enthalpy of each substance at a different temperature, we obtain

$$\Delta H_r(p_0, T) = 3[H_C(p_0, T_0) + C_{p,\,C}(T - T_0)] + [H_D(p_0, T_0) + C_{p,\,D}(T - T_0)]$$
$$- 2[H_A(p_0, T_0) + C_{p,\,A}(T - T_0)] - [H_B(p_0, T_0) + C_{p,\,B}(T - T_0)].$$

Similarly, for the entropy change of the reaction, we start with the standard condition

$$\Delta S_r(p_0, T_0) = 3S_C(p_0, T_0) + S_D(p_0, T_0) - 2S_A(p_0, T_0) - S_B(p_0, T_0).$$

At any temperature T, the entropy change is

$$\Delta S_r(p_0, T) = 3\left[S_C(p_0, T_0) + C_{p,\,C} \ln\left(\frac{T}{T_0}\right) \right] + \left[S_D(p_0, T_0) + C_{p,\,D} \ln\left(\frac{T}{T_0}\right) \right]$$
$$- 2\left[S_A(p_0, T_0) + C_{p,\,A} \ln\left(\frac{T}{T_0}\right) \right] - \left[S_B(p_0, T_0) + C_{p,\,B} \ln\left(\frac{T}{T_0}\right) \right].$$

Finally, the Gibbs free energy of the reaction at standard conditions is

$$\Delta G_r(p_0, T_0) = 3G_C(p_0, T_0) + G_D(p_0, T_0) - 2G_A(p_0, T_0) - G_B(p_0, T_0).$$

At any temperature T, the Gibbs free energy change is

$$\Delta G_r(p_0, T) = 3[H_C(p_0, T) - TS_C(p_0, T)] + [H_D(p_0, T) - TS_D(p_0, T)] - 2[H_A(p_0, T) - TS_A(p_0, T)]$$
$$- [H_B(p_0, T) - TS_B(p_0, T)].$$

Alternatively, this expression simplifies to

$$\Delta G_r(p_0, T) = \Delta H_r(p_0, T) - T\Delta S_r(p_0, T).$$

This simplification provides a more straightforward way to compute the Gibbs free energy change at non-standard temperatures.

9.6.5 Heat capacity of a reaction

The heat capacity of the reaction can be expressed as

$$C_{p,r} = 3C_{p,C} + C_{p,D} - 2C_{p,A} - C_{p,B}.$$

Thus, the enthalpy and entropy of a reaction can be rewritten in terms of the reaction's heat capacity:

$$\Delta H_r(p_0, T) = \Delta H_r(p_0, T_0) + C_{p,r}(T - T_0)$$

$$\Delta S_r(p_0, T) = \Delta S_r(p_0, T_0) + C_{p,r} \ln\left(\frac{T}{T_0}\right).$$

These equations simplify the calculations for reactions under varying temperatures, enabling easier estimation of the thermodynamic properties during the reaction process.

9.7 Determining Gibbs free energy of a reaction at any pressure

Let us consider the same reaction as before but focus on how pressure affects the Gibbs free energy.

In order to account for the pressure dependence, we need to identify the phase of each substance in the reaction—solid (s), liquid (l), or gas (g):

$$2A\ (s) + B\ (g) \longrightarrow 3C\ (g) + D\ (l).$$

For each substance, the change in Gibbs free energy $dG(p, T)$ is given by the equation

$$dG(p, T) = -SdT + Vdp.$$

Since we are interested in the pressure dependence, the temperature term (dT) becomes zero. Therefore, we have

$$dG(p, T) = Vdp.$$

For a reaction, the change in Gibbs free energy $\Delta G_r(p, T)$ is related to the change in volume of the system,

$$d\Delta G_r(p, T) = \Delta V_r dp.$$

The volume change in the reaction, ΔV_r, is the difference in volumes between the products and reactants:

$$\Delta V_r = 3V_C + V_D - 2V_A - V_B.$$

Since the volume of a gas is much larger than that of a solid or liquid, we can generally neglect the volume contributions of solids and liquids when gases are present in a reaction. However, it is important to note that if the reaction involves only solids and liquids, we should not neglect their volumes. For example, in the reaction where graphite converts to diamond (both solids), the volume of both substances must be considered.

Returning to our reaction, we can neglect the volumes of substances A and D, which are solids or liquids. For gases, we use the ideal gas expression for one mole of a substance. Therefore, the volume change in the reaction becomes

$$\Delta V_r = 3RT/p + 0 - 2 \cdot 0 - RT/p = 2RT/p.$$

Now, the change in Gibbs free energy for the reaction at any pressure is

$$d\Delta G_r = 2RT/p \ dp.$$

Integrating this expression gives the total change in Gibbs free energy between the pressures p_0 and p:

$$\Delta G_r(p) - \Delta G_r(p_0) = \int_{p_0}^{p} 2RT/p \ dp = 2RT \ln\left(\frac{p}{p_0}\right).$$

This approach shows how to determine the pressure dependence of Gibbs free energy for a reaction. By understanding the volume change in the reaction and applying the ideal gas law, we can derive an expression for the Gibbs free energy that is applicable at any pressure. This method is essential for understanding how reactions behave under varying pressure conditions.

9.8 Gibbs free energy and reaction behavior under non-standard temperature and pressure

In thermodynamics, understanding how a reaction behaves under various pressure and temperature conditions is critical for predicting its spontaneity and feasibility. To start, it is important to recall the fundamental equation for Gibbs free energy:

$$\Delta G_r = \Delta H_r - T\Delta S_r.$$

For the sake of simplicity in the following analysis, we will assume that the temperature dependence of enthalpy (H) and entropy (S) is negligible. This will allow us to focus on the key relationship between Gibbs free energy and temperature, but it is worth noting that this is a rough approximation. If we plot this equation, with Gibbs free energy (ΔG_r) on the y-axis and temperature on the x-axis, we obtain a straight line. The Y-intercept represents the change in enthalpy (ΔH_r), and the slope is given by $-\Delta S_r$ (see figure 9.3).

This graphical approach is extremely useful because it offers a quick way to predict how reactions behave at different temperatures. Although this method does not provide exact values, it guides us in the right direction and forms a useful first tool for thermodynamic investigations.

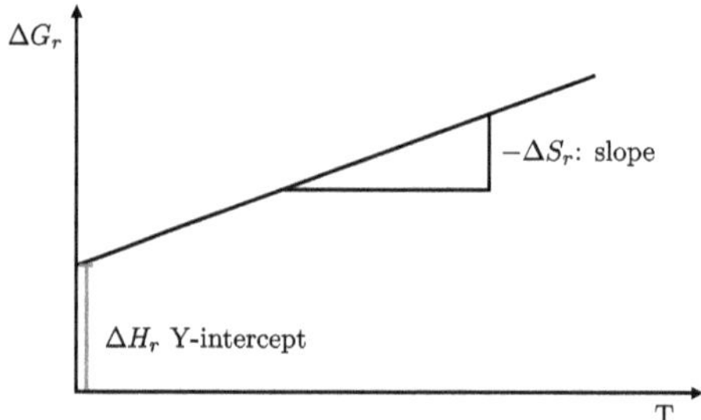

Figure 9.3. Temperature dependence of the Gibbs free energy change of a chemical reaction, ΔG_r, at constant pressure. The plot illustrates the linear relation $\Delta G_r = \Delta H_r - \Delta S_r$, where the y-intercept corresponds to the enthalpy change of reaction ΔH_r, and the slope equals the negative of the entropy change of reaction ΔS_r.

9.8.1 Case studies: reaction behavior under various conditions

Now, let us apply this graphical analysis to a few example reactions to observe how their behavior changes under different pressure and temperature conditions. Each case is an example of a chemical reaction where the entropy and enthalpy changes are different, impacting the spontaneity of the process.

9.8.1.1 Case 1: $\Delta S_r > 0$; $\Delta H_r < 0$
Consider the reaction

$$C\ (s) + O_2\ (g) \longrightarrow CO_2\ (g).$$

The relevant data for the substances involved in this reaction are shown in the table on the left-hand side of figure 9.4.

For this reaction, the entropy change (ΔS_r) is positive, and the enthalpy change (ΔH_r) is negative. This means the reaction is always spontaneous at all temperatures. The slope of the Gibbs free energy versus temperature plot is negative because the entropy is increasing, and the Y-intercept is negative because the reaction is exothermic (releases heat) (see the graph on the right-hand side of figure 9.4).

9.8.1.2 Case 2: $\Delta S_r > 0$; $\Delta H_r > 0$
Now, consider the reaction

$$PCl_5\ (g) \longrightarrow PCl_3\ (g) + Cl_2\ (g).$$

The relevant data for this reaction are shown in the table on the left-hand side of figure 9.5.

In this case, the slope of the ΔG_r versus temperature plot is negative (due to positive entropy), but the Y-intercept is positive because the reaction starts with a positive enthalpy change. The plot crosses the x-axis at 518 K, which is the transition temperature where the process becomes spontaneous. Below this temperature, the reaction is non-spontaneous, while above it, it becomes spontaneous (see the graph on the right-hand side of figure 9.5).

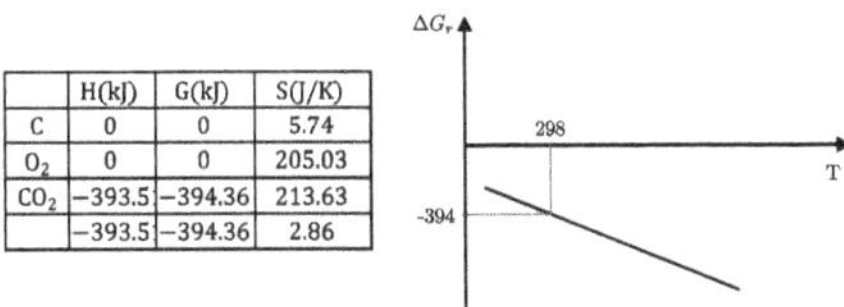

	H(kJ)	G(kJ)	S(J/K)
C	0	0	5.74
O_2	0	0	205.03
CO_2	−393.5	−394.36	213.63
	−393.5	−394.36	2.86

Figure 9.4. Temperature dependence of the Gibbs free energy change for the formation of carbon dioxide. The plot is based on standard thermodynamic data shown in the table. The negative slope indicates that the reaction becomes more favorable at higher temperatures.

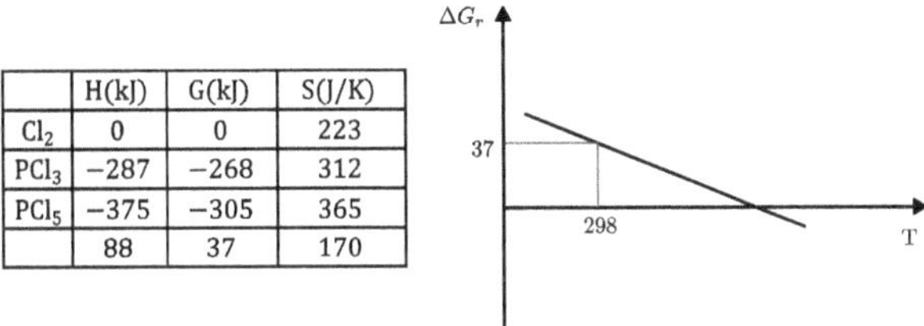

	H(kJ)	G(kJ)	S(J/K)
Cl_2	0	0	223
PCl_3	−287	−268	312
PCl_5	−375	−305	365
	88	37	170

Figure 9.5. Gibbs free energy change versus temperature for the formation of phosphorus pentachloride (PCl_5) from phosphorus trichloride (PCl_3) and chlorine (Cl_2), based on standard thermodynamic data. The reaction becomes spontaneous above a certain temperature.

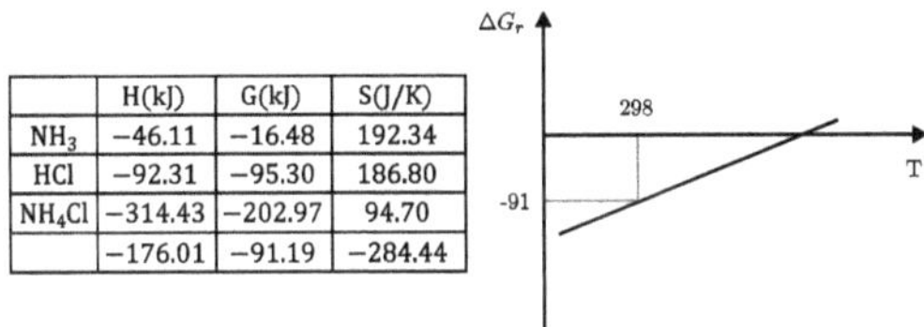

	H(kJ)	G(kJ)	S(J/K)
NH_3	−46.11	−16.48	192.34
HCl	−92.31	−95.30	186.80
NH_4Cl	−314.43	−202.97	94.70
	−176.01	−91.19	−284.44

Figure 9.6. Temperature dependence of the Gibbs free energy change for the formation of ammonium chloride (NH_4Cl) from ammonia (NH_3) and hydrogen chloride (HCl), based on standard thermodynamic data.

9.8.1.3 Case 3: $\Delta S_r < 0$; $\Delta H_r < 0$

Consider the reaction

$$NH_3\,(g) + HCl\,(g) \longrightarrow NH_4Cl\,(s).$$

The data for this reaction are shown in the table on the left-hand side of figure 9.6.

In this case, the reaction has a negative entropy change ($\Delta S_r < 0$) and a negative enthalpy change ($\Delta H_r < 0$). The curve has a positive slope because the entropy is negative. The reaction is initially spontaneous at lower temperatures, but as the temperature increases, the system reaches a transition point around 619 K, beyond which the reaction becomes non-spontaneous (see the graph on the right-hand of figure 9.6).

9.8.1.4 Case 4: $\Delta S_r < 0$; $\Delta H_r > 0$

Consider the reaction

$$C\,(graphite) \longrightarrow C\,(diamond).$$

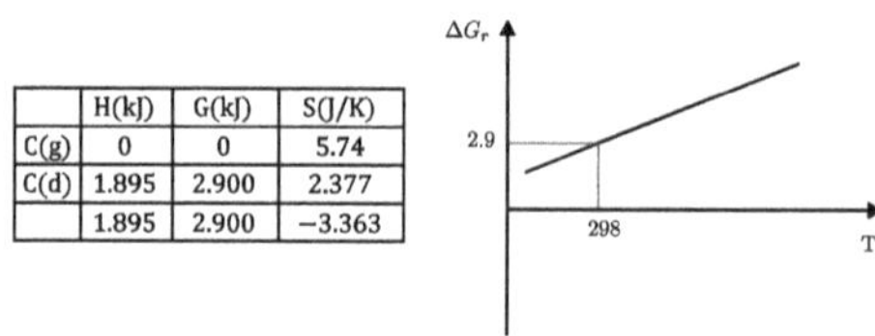

Figure 9.7. Temperature dependence of the Gibbs free energy change for the conversion of graphite (C(g)) to diamond (C(d)), illustrating the thermodynamic favorability of graphite at room temperature.

The data for this reaction are shown in the table on the left-hand side of figure 9.7.

This reaction involves the transformation of graphite into diamond. As the reaction involves a negative entropy change ($\Delta S_r < 0$) and a positive enthalpy change ($\Delta H_r > 0$), the reaction will never occur spontaneously at any temperature. The transition from graphite to diamond is not driven by temperature but may be influenced by extreme pressure conditions, which we will discuss in the next section.

9.8.2 Temperature dependence of enthalpy and entropy

While the graphical analysis above provides a useful first step in understanding the behavior of reactions at different temperatures, we have neglected the temperature dependence of enthalpy and entropy. By incorporating this dependency, we can obtain a better approximation of the system's behavior, though the result remains an estimate due to the assumption of constant heat capacities.

For example, in case 3, we previously calculated the enthalpy and entropy at the transition temperature of 619 K. Including the temperature dependence of heat capacities yields more accurate values for enthalpy and entropy at this temperature, refining our estimate of the Gibbs free energy,

$$\Delta G_r = \Delta H_r - T\Delta S_r.$$

This approach is still an approximation because we assume heat capacities are constant, but it provides more accurate predictions of the reaction's behavior at different temperatures.

9.8.3 Example with hydrogen

Hydrogen is a key player in the ongoing energy transition and is of particular interest for various energy applications. To illustrate the concepts of Gibbs free energy and its implications, we will examine hydrogen in several reactions, considering three main cases: combustion, fuel cells, and electrolysis.

9.8.4 Combustion of hydrogen

The combustion of hydrogen is a straightforward and important reaction:

$$H_2\,(g) + 0.5O_2\,(g) \longrightarrow H_2O\,(l).$$

The thermodynamic data for this reaction are shown in table 9.2.

Table 9.2. Standard enthalpy, Gibbs free energy, and entropy values for hydrogen, oxygen, and water at 298 K.

Substance	H (kJ)	G (kJ)	S (J K^{-1})
H_2	0.00	0.00	130.68
O_2	0.00	0.00	102.57
H_2O	−285.83	−237.13	69.91

The Gibbs free energy for the reaction can be calculated using the equation

$$\Delta G_r = \Delta H_r - T\Delta S_r.$$

Substituting the values,

$$\Delta G_r = -285.83 - (298 \times (-163.34 \text{ J K}^{-1})).$$

This results in

$$\Delta G_r = -237.16 \text{ kJ mol}^{-1} \text{ of } H_2.$$

It is important to note that this Gibbs free energy is calculated per mole of hydrogen or half mole of oxygen or one mole of water. If you are asked to calculate the Gibbs free energy per mole of oxygen, you need to multiply the value by two, which gives −474.32 kJ mol^{-1} of O_2.

This reaction falls under case 3 from earlier discussions, where the process is spontaneous at room temperature. However, this reaction will have a **transition temperature** where the process becomes non-spontaneous. In this case, that temperature would be around 1750 K. The question arises: why does hydrogen still burn on the Sun despite the high temperature? The answer lies in the extremely high pressure on the Sun's surface (although the Sun technically does not have a solid surface, its outer layers exhibit high pressure), which significantly increases the transition temperature.

In this case, the work done by the system (W_{other}) is zero, meaning that all the energy released is in the form of heat. Therefore, the heat transfer is

$$Q = \Delta H_r - W_{\text{other}} = -285.83 \text{ kJ mol}^{-1} \text{ of } H_2.$$

This is an exothermic reaction, which releases energy. A famous (albeit dangerous) application of this reaction is hydrogen-filled balloons, as seen in the Zeppelin disaster—hydrogen's flammability can be highly destructive.

9.8.5 Hydrogen fuel cell

The hydrogen fuel cell operates on a similar principle to hydrogen combustion but with an important difference: it generates electrical energy by transporting electrons through a circuit. The reaction in a fuel cell is

$$H_2 \text{ (g)} + 0.5O_2 \text{ (g)} + 2e^- \longrightarrow H_2O \text{ (l)} + 2e^-.$$

Here, we have the same chemical reaction, but it also involves the movement of electrons, producing work. The energy output is captured as electrical work, and the maximum work $|W_{other}|$ is equal to the Gibbs free energy:

$$|W_{other}| = \Delta G_r = 237.16 \, \text{kJ mol}^{-1} \text{ of } H_2.$$

This is a **negative work** value, meaning the system is delivering energy to the surroundings in the form of electrical work.

The heat transfer in this case is

$$Q = T\Delta S_r = -48.67 \, \text{kJ mol}^{-1} \text{ of } H_2.$$

This indicates that the reaction is exothermic.

What happens if the fuel cell operates at a higher pressure, say 100 bar? The change in Gibbs free energy with pressure can be determined by the volume change of the reaction

$$d\Delta G_r = \Delta V_r dp.$$

The volume change for the reaction is given by

$$\Delta V_r = V_{H_2O} - V_{H_2} - 0.5 V_{O_2}.$$

Since V_{H_2O}, the volume of liquid water, is negligible, we use the ideal gas law to approximate the volume change of hydrogen and oxygen. The volume change is then

$$\Delta V_r = -1.5 \cdot RTp.$$

Integrating this expression gives

$$\Delta G_r(p) - \Delta G_r(p_0) = \int_{p_0}^{p} 1.5RT \ln\left(\frac{p}{p_0}\right) dp.$$

At 100 bar, the Gibbs free energy is

$$\Delta G_r(p = 100 \, \text{bar}) = -254 \, \text{kJ mol}^{-1}.$$

Since the absolute value of ΔG_r is larger at 100 bar compared to 1 bar, the work output of the fuel cell is larger, resulting in higher voltage.

9.8.6 Hydrogen electrolysis

Hydrogen electrolysis is the reverse of the fuel cell reaction, where water is split into hydrogen and oxygen. The reaction is

$$H_2O + 2e^- \longrightarrow H_2 \, (g) + 0.5O_2 \, (g) + 2e^-.$$

This is a non-spontaneous reaction at room temperature, meaning it requires external work to proceed. The minimum work required is

$$\Delta G_r = 237.16 \, \text{kJ mol}^{-1} \text{ of } H_2.$$

This shows that electrolysis requires energy input, which is why we use electrical power to split water into hydrogen and oxygen.

9.8.7 Hydrogen combustion at higher pressure

When hydrogen combusts at higher pressures, the Gibbs free energy changes with pressure. As pressure increases, the transition temperature at which the reaction changes from spontaneous to non-spontaneous also increases. At extremely high pressures, such as those on the Sun's surface (estimated at 265 billion bar), the transition temperature becomes very large, making hydrogen combustion feasible even at temperatures far beyond Earth's typical conditions.

9.9 Understanding water's phase diagram

The principles we have applied to chemical reactions can also be used for physical phase changes. Let us take the example of the phase change of water from liquid to vapor (boiling). The reaction is

$$H_2O \ (l) \longrightarrow H_2O \ (g).$$

The Gibbs free energy change for this process determines the boiling point of water at a given pressure. Using data for enthalpy and entropy at standard conditions, we can calculate the transition temperature or boiling point. For water, this transition temperature is typically around 373.13 K at 1 bar, but it can be influenced by pressure.

By applying the Gibbs free energy equation,

$$\Delta G_r = \Delta H_r - T \Delta S_r.$$

We can analyse the boiling point of water at various pressures. For example, if we calculate the Gibbs free energy at 370.08 K, we find $\Delta G_r = -1.55$ kJ, which still slightly deviates from the expected value due to the abrupt change in heat capacity near the boiling point. However, using the data at 298 K and solving for the pressure at which water boils, we find

$$p = 0.032 \ \text{bar}.$$

This shows that water will boil at 298 K at a much lower pressure than the standard boiling point.

This analysis of hydrogen and water's behavior under varying pressures and temperatures shows how the principles of Gibbs free energy can be applied not only to chemical reactions but also to physical phase changes. Understanding these concepts is essential for predicting reaction spontaneity and studying phase transitions, both of which are crucial in a wide range of industrial and scientific applications.

9.10 Calculating electromotive force (EMF) in electrochemical cells

The electromotive force (EMF) of an electrochemical cell represents the maximum potential difference between its two electrodes when no current is flowing. This potential difference drives the movement of electrons through an external circuit, facilitating the electrochemical reaction. Understanding how to calculate EMF is essential for analysing and designing electrochemical systems such as batteries and fuel cells.

9.10.1 Fundamental relationship between Gibbs free energy and EMF

The relationship between the Gibbs free energy change (ΔG) of a reaction and the EMF of an electrochemical cell is given by

$$\Delta G = -nFE_{\text{cell}},$$

where:
- ΔG is the Gibbs free energy change (in joules).
- n is the number of moles of electrons transferred in the reaction.
- F is the Faraday constant (96485 C mol^{-1}).
- E_{cell} is the EMF of the cell (in volts).

This equation indicates that the EMF of a cell is directly related to the spontaneity of the electrochemical reaction: a positive EMF corresponds to a negative ΔG, signifying a spontaneous reaction.

9.10.1.1 Step 1: Calculate the Gibbs free energy change for the reaction
First, determine the standard Gibbs free energy change (ΔG°) for the reaction. This is typically done using standard thermodynamic data (enthalpy and entropy) for the substances involved.

The standard Gibbs free energy change is given by

$$\Delta G^\circ = \Delta H^\circ - T\Delta S^\circ,$$

where:
- ΔH° is the standard enthalpy change (in joules, J).
- T is the temperature (in kelvins, K).
- ΔS° is the standard entropy change (in joules per kelvin, $J\ K^{-1}$).

9.10.1.2 Step 2: Relate Gibbs free energy to EMF
Once you have the standard Gibbs free energy change ΔG° for the reaction, use the equation relating Gibbs free energy to EMF:

$$\Delta G^\circ = -nFE^\circ_{\text{cell}}.$$

Rearrange this equation to solve for the EMF:

$$E^\circ_{\text{cell}} = -\frac{\Delta G^\circ}{nF}.$$

9.11 Pressure dependence of the graphite–diamond transition

In chapter 8, we discussed the transformation from graphite to diamond under standard conditions, where the Gibbs free energy change (ΔG) was positive, indicating that the transformation is non-spontaneous at 1 atm pressure and room temperature. However, pressure plays a significant role in this phase transition. At higher pressures, the reaction becomes more favorable, and the transition from graphite to diamond can occur spontaneously.

The formula we will use to describe the pressure dependence of this reaction is

$$d\Delta G_\gamma = \Delta V_\gamma\, dp - \Delta S_\gamma\, dT.$$

9.11.1 Step 1: Understanding the volume and entropy change

For the graphite to diamond transformation, the volume change (ΔV_γ) is negative, as diamond is denser than graphite. Therefore, the volume change ΔV_γ is a key factor in driving the reaction forward under high pressure.

The entropy change (ΔS_γ) for the reaction is negative because diamond has a more ordered structure than graphite. However, at high pressures, the contribution from the $\Delta_\gamma dp$ term becomes much more significant and helps drive the reaction forward, making the transformation spontaneous.

9.11.2 Step 2: Calculating Gibbs free energy change at standard pressure

At standard conditions (298 K and 1 atm), we can calculate the Gibbs free energy change for the transformation from graphite to diamond. From earlier discussions, we know that

$$\Delta H_\gamma = 1.895 \text{ kJ mol}^{-1}, \quad \Delta S_\gamma = -3.363 \text{ J K}^{-1}\text{·mol}^{-1}.$$

The Gibbs free energy change at standard pressure and temperature can be calculated as

$$\Delta G_\gamma = \Delta H_\gamma - T\Delta S_\gamma.$$

Substituting the values,

$$\Delta G_\gamma = 1.895 \text{ kJ mol}^{-1} - \left(298 \text{ K} \times \left(-3.363 \text{ J} \quad \text{K}^{-1}\text{·mol}^{-1} \times \frac{1}{1000 \text{ J kJ}^{-1}}\right)\right)$$

$$\Delta G_\gamma = 1.895 \text{ kJ mol}^{-1} + 1.006 \text{ kJ mol}^{-1}$$

$$\Delta G_\gamma = 2.901 \text{ kJ mol}^{-1}.$$

This positive value shows that the transformation is non-spontaneous at standard conditions.

9.11.3 Step 3: Calculating Gibbs free energy change at high pressure

Now, we consider a high-pressure environment, such as 5 GPa, which is typical for synthetic diamond production. We will calculate how the pressure dependence affects the Gibbs free energy change.

9.11.3.1 Volume change and pressure contribution

The volume change for the graphite to diamond transformation is approximately

$$\Delta V_\gamma \approx -5.7 \times 10^{-6} \text{ m}^3 \text{ mol} - 1.$$

At a pressure of 5 GPa, the pressure contribution to the Gibbs free energy change is

$$\Delta V_\gamma \Delta p = (-5.7 \times 10^{-6}\,\text{m}^3\,\text{mol}^{-1}) \times (5 \times 10^9\,\text{Pa})$$

$$\Delta V_\gamma \Delta p = -28.5\,\text{kJ}\,\text{mol}^{-1}.$$

9.11.3.2 Temperature contribution

Since we are assuming room temperature ($T = 298$ K), and there is no significant change in temperature ($\Delta T = 0$), the temperature term in the equation is zero:

$$\Delta T = 0.$$

9.11.3.3 Total Gibbs free energy change

Now, using the pressure contribution and the earlier values for enthalpy and entropy, we calculate the total Gibbs free energy change at high pressure:

$$\Delta G_\gamma = 1.895\,\text{kJ}\,\text{mol}^{-1} - 1.006\,\text{kJ}\,\text{mol}^{-1} + (-28.5\,\text{kJ}\,\text{mol}^{-1})$$

$$\Delta G_\gamma = -27.611\,\text{kJ}\,\text{mol}^{-1}.$$

At 5 GPa, the Gibbs free energy change is negative, indicating that the graphite to diamond transformation is *spontaneous* at this pressure.

This analysis shows that while the graphite to diamond transformation is non-spontaneous at standard pressure, it becomes spontaneous at high pressures due to the significant contribution of the volume change term $\Delta V_\gamma \Delta p$ in the Gibbs free energy equation. This effect is crucial in the synthetic production of diamonds, where high pressures are used to drive the reaction. By using the correct thermodynamic equation, we can predict that the transformation from graphite to diamond becomes favorable under high pressure, and this provides insight into the conditions needed for diamond synthesis in both natural and industrial settings.

9.12 Summary

In this chapter, we explored the central role of thermodynamic potentials in understanding the behavior of systems under various conditions. The relationships between internal energy, enthalpy, and Gibbs free energy were discussed in detail, with emphasis on their mathematical formulations and physical significance.

We demonstrated how the first and second laws of thermodynamics can be applied to calculate energy changes and work done in various systems. The chapter addressed work expressions, showing how they differ in systems undergoing compression and expansion (using $-pdV$ work) versus those involving flow processes (using Vdp work). By clearly distinguishing these work terms, we emphasized their relevance to different practical processes, including engines, compressors, and refrigerators.

A significant portion of the chapter was dedicated to examining the pressure and temperature dependence of reactions. We considered the transformation of graphite to diamond as a case study, demonstrating how pressure influences the spontaneity of phase transitions. The use of Gibbs free energy and its relationship with pressure allowed us to predict when such transformations become favorable.

The latter sections of the chapter delved into the methods for calculating enthalpy, entropy, and Gibbs free energy at non-standard pressures and temperatures. We extended these concepts to reactions and phase changes, such as the combustion of hydrogen and the boiling of water, emphasizing how these properties evolve with changes in environmental conditions.

Finally, we explored the application of these principles to electrochemical cells, particularly focusing on the calculation of electromotive force (EMF) and its dependence on Gibbs free energy. We highlighted how understanding these relationships can aid in the design of efficient energy systems, such as fuel cells and batteries, which are essential for modern technological advancements.

In summary, this chapter provided a thorough understanding of thermodynamic potentials, their derivatives, and how to apply them to real-world scenarios. Through examples and case studies, we illustrated how these principles guide us in predicting the behavior of substances and reactions under varying temperature and pressure conditions. The knowledge gained here is vital for anyone involved in fields that deal with energy transformations, chemical processes, or material properties under different environmental constraints.

Chapter 10

Conclusion: where to go from here

Thermodynamics begins simply—with energy, temperature, and pressure—but quickly reveals itself to be a universal language. In this book, we have walked through that language together, chapter by chapter, building a foundation rooted in intuition and clarity.

We started with energy—the mighty internal energy U—the strength within a system. Then we brought in entropy S, which asks: *How is that energy arranged?* From there, we met enthalpy H, which accounts for the energy needed not just to grow, but to push back the world—a concept essential in systems open to the atmosphere. Finally, we encountered the Gibbs free energy G, the portion of energy available to do useful work when all other requirements are met.

To conclude, consider this playful analogy with elephants (figure 10.1)—each one representing a thermodynamic potential:

- The **first elephant** stands tall, full of internal energy U.
- The **second elephant**, now grown and pushing against the air, needs more—the total $H = U + pV$.
- The **third elephant**, destined for transformation, shows that the maximum useful work is not all of H, but only the portion we call G.

With this journey, we've not only met these characters but followed them through processes, cycles, and phase transitions. We have seen that thermodynamics is not confined to heat engines or classical systems. It applies just as much to modern technologies—in microelectronics, chemistry, renewable energy, and even quantum systems.

Yet, this is not the end. Some essential topics lie beyond the scope of this book, particularly those that are crucial in the context of sustainability and energy systems. Concepts like **availability, exergy,** and **irreversibility accounting** are necessary tools to evaluate the *quality* of energy and to determine how much can truly be harnessed

doi:10.1088/978-0-7503-6029-6ch10

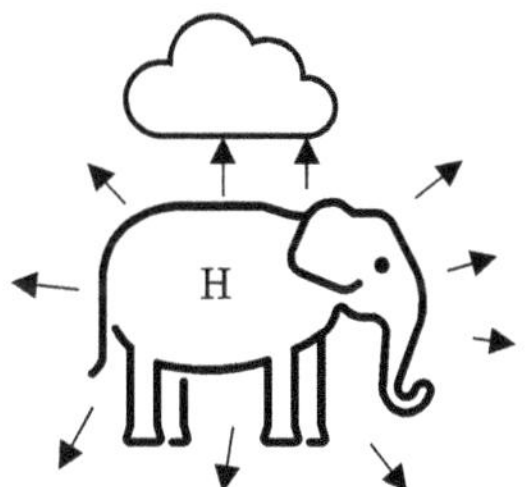

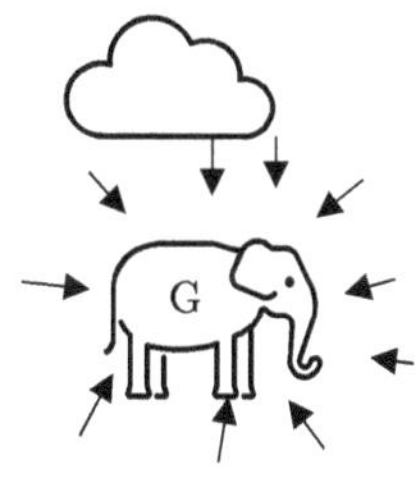

Figure 10.1. A playful analogy using elephants to illustrate thermodynamic potentials: the first elephant represents internal energy U, the second shows enthalpy H, accounting for work done against atmospheric pressure, and the third illustrates Gibbs free energy G, the portion of energy available to do useful work.

before it is lost to entropy. These ideas are foundational to next-generation thinking in energy system design and deserve a dedicated study.

For deeper explorations, classic textbooks such as Van Wylen and Sonntag or Moran and Shapiro offer rigorous mathematical treatments. They are valuable companions as you move from this conceptual introduction into more advanced or domain-specific applications—whether in mechanical engineering, chemical thermodynamics, or materials science.

The next step is up to you. Perhaps you will apply these ideas in designing cryogenic systems, optimizing fuel cells, modeling climate interactions, or even interpreting biological processes. No matter the direction, the core principles remain the same—powerful, universal, and deeply insightful.

Let this be your springboard. The tools are in your hands. The elephants have spoken.

Further reading

The following books provide complementary perspectives on thermodynamics and its applications. Each emphasizes different aspects of the subject, from conceptual clarity to mathematical rigor, and from classical foundations to modern cryogenic practice. Readers are encouraged to consult them for deeper study and alternative explanations of key ideas introduced in this book.

An Introduction to Thermal Physics, Daniel V Schroeder, 2021
A beautifully written text for physics students, offering an accessible and intuitive introduction to energy, entropy, and temperature. Its greatest strength lies in the statistical interpretation of entropy and the seamless connection between microscopic and macroscopic descriptions. However, the treatment of open systems, other forms of work, and thermodynamic cycles is limited. Topics such as irreversible and

quasistatic processes or spontaneity at non-standard conditions are not explored in depth, but the book remains one of the clearest conceptual text to thermal physics.

Fundamentals of Thermodynamics (SI Version, 7th Edition), Claus Borgnakke and Richard E Sonntag
A cornerstone of engineering thermodynamics education, this text provides a comprehensive introduction supported by numerous worked examples and problems. It follows the engineering sign convention (heat in and work out are positive). The book is particularly strong in its systematic development of properties, cycles, and energy balances. While its focus is primarily on closed systems and engineering processes, discussions of non-quasistatic processes and spontaneity at non-standard conditions are limited. Nevertheless, it remains an indispensable companion for anyone seeking an engineering-oriented approach to classical thermodynamics.

Concepts in Thermal Physics, Stephen J Blundell and Katherine M Blundell
A modern and engaging text that bridges thermodynamics and statistical mechanics, written with clarity for physics students. It covers advanced topics such as Fermi–Dirac and Bose–Einstein statistics, phase transitions, and quantum effects in matter. However, it does not devote much attention to the thermodynamic treatment of irreversibility, other work interactions, or spontaneous processes, aspects emphasized in this book. It serves as an excellent second text for readers wishing to extend their understanding beyond classical thermodynamics.

Fundamentals of Thermal-Fluid Sciences, Yunus A Çengel, John M Cimbala, and Afshin J Ghajar
This book combines thermodynamics, fluid mechanics, and heat transfer, offering a unified perspective for engineering students. It provides an extensive range of examples and exercises that connect theory with practice. The coverage of energy conservation and cycle analysis is broad, though the conceptual discussions on irreversibility and spontaneity remain brief. As an introductory resource, it is particularly valuable for those wishing to see how thermodynamic principles extend to transport and flow processes.

Cryogenic Helium Refrigeration for Middle and Large Powers, Guy Gistau, 2019
For readers interested in low-temperature engineering, this book provides a detailed overview of helium-based cryogenic refrigeration systems used in large-scale applications such as accelerators, fusion reactors, and space missions. It introduces advanced thermodynamic cycles, component-level descriptions, and system integration methods. The text blends theoretical analysis with practical engineering experience, making it an essential reference for those venturing into applied cryogenics.

Helium Cryogenics, Steven W Van Sciver, 2nd Edition, Springer, 2012
A classic reference that captures both the classical and quantum aspects of helium, the only element that remains liquid down to absolute zero at normal pressures. The book

combines detailed thermophysical property data with rigorous analyses of refrigeration and liquefaction systems. It illustrates how helium behaves as a bridge between macroscopic thermodynamics and quantum phenomena, covering subjects such as superfluidity, heat transport, and cryogenic component design. Together with *The Handbook of Cryogenic Engineering* (edited by J Weisend II), it forms a foundational resource for advanced students and professionals in low-temperature science and engineering.

www.ingramcontent.com/pod-product-compliance
Ingram Content Group UK Ltd.
Pitfield, Milton Keynes, MK11 3LW, UK
UKHW051938150726
7214IPUK00005B/40